# Firms in Open Source Software Development

Mario Schaarschmidt

# Firms in Open Source Software Development

## Managing Innovation Beyond Firm Boundaries

Foreword by Prof. Dr. Harald von Kortzfleisch

 Springer Gabler          **RESEARCH**

Mario Schaarschmidt
University of Koblenz-Landau, Germany

Vollständiger Abdruck der von der Universität Koblenz-Landau genehmigten Dissertation.

ISBN 978-3-8349-4142-8     ISBN 978-3-8349-4143-5 (eBook)
DOI 10.1007/978-3-8349-4143-5

The Deutsche Nationalbibliothek lists this publication in the Deutsche Nationalbibliografie; detailed bibliographic data are available in the Internet at http://dnb.d-nb.de.

Springer Gabler
© Gabler Verlag | Springer Fachmedien Wiesbaden 2012

*Cover design:* KünkelLopka GmbH, Heidelberg

Printed on acid-free paper

Springer Gabler is a brand of Springer DE. Springer DE is part of Springer Science+Business Media.
www.springer-gabler.de

# Foreword

Since Henry Chesbrough introduced his book Open Innovation in 2003, many researchers have contributed to our understanding of innovation as a phenomenon not soly domiciled within the boundaries of the firm. In particular, both the integration of external sources into corporate innovation processes as well as the external exploitation of innovations generated within the boundaries of the firm have been investigated. Prevalent in this discussion seems to be that (1) external resources can complement the firm's own resource base and (2) integration of these resources may lead to increased innovative perfomance.

As the example of open source software furher shows, firms not necessarily have to pay for such external resources to obtain. In these scenarios, individuals provide their knowledge for free, by, for example, reporting bugs, requesting new features or - at least in smaller projects - contributing pieces of code.

However, this advantage of getting access to external resources comes at larger coordination and control costs that may outperfom the benefits. For instance, in the case of open source software, the number of participants could - theoretically - be very large which makes it difficult for a firm to influence a project's trajectory. Yet, *how* firms control what happens beyond their boundaries - and beyond their vertical command chain - is outside of our knowledge.

With this book, Mario Schaarschmidt contributes to the understanding of the open innovation phenomenon by focusing on a part that has not received much attention yet: managaging, and therefore pursuing control, beyond the boundaries of the firm.

This book is Mario Schaarschmidt's doctoral thesis at University of Koblenz-Landau and depicts a starting point of a fruitful academic career. While wishing the readers the same interesting insights I had when reading Mario's dissertation, I can recommend it to academics and practitioners alike.

*Prof. Dr. Harald von Kortzfleisch*
Koblenz, Germany

# Preface

I would like to take this space to thank all those who made this dissertation possible. First and foremost, I have to thank Harald von Kortzfleisch, my thesis advisor, for giving me the freedom to be creative within my thesis and for providing me with necessary (financial) resources to become a better researcher. Harald, without your support, this thesis never could have been done. Second, I thank Gianfranco Walsh for taking over the part as a second supervisor and for his British humor which helped me to survive dark hours during the completion of this thesis.

Furthermore, I want to thank my parents, Ursula and Rainer, who took over lots of inconvenient tasks in everyday life so I was able to concentrate on this thesis. Same holds true for my Grandma and my sister, Ute.

I further like to warmly thank people at the University of Koblenz-Landau, in particular Thomas Kilian, Berthold Hass, Matthias Bertram, and Nadine Lindermann, for numerous valuable discussions and suggestions. A number of great researchers guided my way to this dissertation. While working as a student at WHU - Otto Beisheim School, I have been inspired by Ulrich Lichtenthaler. Ines Mergel hosted me at Harvard's Kennedy School and gave me a wonderful opportunity to enrich my knowledge base. Finally, I have to thank Linus Dahlander for the work I could build upon as well as Ricco Deutscher for an invaluable interview on open source strategies.

I also like to thank participants of several conferences and workshops I attended for valuable comments of parts of this thesis and for broadening my scope. These meetings were, amongst others, various European Academy of Management Conferences (EURAM), the 8th User and Open Innovation Workshop, MIT Sloan School of Management, Cambridge, MA, the Paper Development Workshop, Academy of Management Annual Meeting (AOM), OCIS Division, Montréal, Canada, and the PhD Seminar on Open Innovation at ESADE Business School, Barcelona, Spain.

For support in generating this thesis I would further like to thank Christoph Schneider whose Master Thesis provided valuable input to parts of the thesis and the 2008 research project team on open source software for collecting data in one case.

Finally, I would like to thank my girlfriend Nadine for her lenience and support.

*Mario Schaarschmidt*
Koblenz, Germany

# Contents

# List of Tables

# List of Figures

# Chapter 1

# Introduction

## 1.1 Motivation

In the software market, the recent trend seems to be moving toward the development, use, and adoption of open source software (OSS). Various venture capital (VC) deals and acquisitions of OSS firms by large software vendors document the increasing prevalence of OSS in business (Von Kortzfleisch et al. 2010), such as the acquisition of JBoss by RedHat in April 2006 for approximately $ 350 million,[1] the acquisition of Sleepycat by Oracle in February 2006[2] or the $ 1 billion acquisition of MySQL by Sun in January 2008.[3] In addition, IBM's investment in Linux (Iansiti & Richards 2006), Google's initiation of the Android platform, which has already reached second place in the mobile operating systems market,[4] or the recently announced alliance between Nokia and Intel for the development of their own OSS operating system for mobile phones named MeGoo[5] show, that even software user firms – and not software vendors only – invest in the OSS development approach.

The remarkable characteristics of OSS are rooted in a set of principles that contradicts those of proprietary software vendors, such as demanding license fees for the use of a prod-

---

[1] "JBoss acquired by RedHat", http://www.theserverside.com/news/thread.tss?thread_id=39866, last access: 4/4/2011

[2] "Oracle and Sleepycat", http://www.oracle.com/us/corporate/Acquisitions/sleepycat/index.html, last access: 4/4/2011, Acquisition price not published

[3] "Der nächste Deal: Sun übernimmt MySQL", http://www.computerwoche.de/nachrichtenarchiv/185 2764/, last access: 4/4/2011

[4] "Android No. 2 Mobile OS: Apple Eats Its Dust", http://www.pcworld.com/article/210384/android _no_2_mobile_os_apple_eats_its_dust.html, last access 4/4/2011

[5] "Allianz der Riesen: Nokia und Intel schließen Open-Source-Bündnis", http://www.spiegel.de/netzwelt/games/0,1518,678001,00.html, last access: 4/4/2011

uct (Andersen & Konzelmann 2008, Chen, Iyigun & Maskus 2007). Conversely, because
OSS is developed by a distributed group of individual programmers interacting via electronic mailing lists rather than within the boundaries of a software vendor, it is free of
charge and therefore differs from proprietary software in the way it is produced and distributed (Ghosh 2005, Kogut & Metiu 2001, Lakhani & Von Hippel 2003, Raymond 1998).

Another prominent characteristic of OSS is that the source code, written in a human
readable programming language, is open to anyone and therefore enables capable users
(which may be individuals or firms) to modify the code according to their own needs
(Von Krogh & Spaeth 2007, Von Hippel & Von Krogh 2003). If a modification contributes
to the quality of the original piece of software, such as fixing a bug or adding new functionalities, users are often willing to give the extension back to the open source software
development project (OSSDP) for reputational reasons (Roberts, Hann & Slaughter 2006,
Shah 2006, Xu, Jones & Shao 2009) or as a form of gift exchange (Bergquist & Ljungberg
2001). As a consequence, usually an OSSDP is surrounded by a relatively heterogeneous
community consisting of developers, bug fixers, users, and, if the project is of commercial
interest, firms (Dahlander & Magnusson 2005).

Recent research confirms that OSSDPs of commercial interest exist in many different
ways according to their revenue model, type of license, development style, number of participating firms, number of participating volunteers, or governance mode (e.g., Bonaccorsi,
Giannangeli & Rossi 2006, Dahlander & Magnusson 2008, West 2003). For example, relevant revenue models range from dual licensing approaches, where a product is offered under
two licenses, one OSS license and (at least) one proprietary license, to approaches in which
the revenue stream entirely is generated through the sale of complementary products or
services (Alexy 2009, Fitzgerald 2006, Olson 2005). Depending on the underlying business
model, firms benefit from an engagement in the development of OSS by getting access
to external knowledge, by reducing costs, or by speeding up the diffusion of a technology
(Ågerfalk & Fitzgerald 2008, Bonaccorsi & Rossi 2006, West & Gallagher 2006).

Building upon organizational structures found in OSSDPs allows firms to obtain resources external to the firm. This reflects a dominant view in modern innovation management research known as the open innovation paradigm (Chesbrough 2003d, Chesbrough,
Vanhaverbeke & West 2006). Firms that apply an OSS approach therefore benefit from
opening up their own proprietary software projects or engaging in existing OSSDPs.

Thus, if a firm is able to encourage participation of external parties, it can reduce development costs by including the external contributions that the firm would otherwise have to pay for.[6] As an OSS product is free of charge and open to anyone, it possesses characteristics of a public good such as nonexcludability and nonrivalry (Von Hippel & Von Krogh 2003, Stürmer, Spaeth & Von Krogh 2009). Therefore, it is almost impossible for those who created the value – a group of volunteers or a group of firms alike – to stop competitors from selling added value (Baldwin & Clark 2006, Dahlander 2005, Lerner & Tirole 2002). Consequently, as natural barriers and intellectual property (IP) protection systems are missing, competition is likely to shift to complementary markets (Parker & Van Alstyne 2005).

Firms acting in a closed innovation environment usually respond to competition by increasing the level of IP protection and building stronger barriers around their own innovations (Bogers 2011, Chesbrough 2003b, Chiaroni, Chiesa & Frattini 2010, MacCormack & Iansiti 2009, Pisano 1990). With OSS, this is not possible as a firm either does not own all the copyright required to build strong barriers, or, in the case in which a firm owns the entire copyright of an OSSDP, an increased IP protection would abolish the benefits of OSS, such as rapid diffusion (Morgan & Finnegan 2008). In a similar vein, based on a common understanding that only few innovations yield value on stand-alone basis, various researchers have pointed to the fact, that keeping innovation closed might not be the best path to capturing value (e.g., Chesbrough 2003b, Parker & Van Alstyne 2005, Pisano & Teece 2007, Teece 2010b).

Building upon the distinction between value creation and value capture (Chatain & Zemsky 2011, Lepak, Smith & Taylor 2007, Narayanan, Yang & Zahra 2009, West 2007), value capture becomes increasingly difficult if other entities control required elements for value creation. In other words, if a firm is willing to build a business model upon an OSSDP, due to the heterogeneous group of stakeholders (i.e., volunteers, other firms), influencing a project's trajectory is disproportionately difficult compared to software development within the boundaries of a firm.

As a consequence of decreasing license fees in the software market, OSS development approaches have become a viable alternative to proprietary approaches to software de-

---

[6]It is worth noting that in the case of firm-initiated OSSDPs, no community – neither user nor developer community –, usually exists from the beginning. Consequently, in some cases, the collective singular "the community" consists of only a few individuals. Getting access to thousands of valuable developers working for free by initiating an OSSDP remains an anecdotal myth (Goldman & Gabriel 2005). Rather, creating an active community in support of an OSSDP might be one of the biggest challenges for firms trying to benefit from engaging in OSS development by initiating their own projects.

velopment and distribution (Augustin 2008, Lerner & Schankerman 2010). However, although harnessing free external resources in a firm-driven software development approach potentially increases the firm's innovative performance and reduces its development costs, without appropriate governance mechanisms, the diverse views on a project's trajectory present in an OSSDP lead to divergence (Almirall & Casadesus-Masanell 2010), resulting in increased coordination and control costs or a fork in the worst case.

Therefore, as firms increasingly deploy resources into OSSDPs (Fosfuri, Giarratana & Luzzi 2008, West & O'Mahony 2008), they need to gain a better understanding of the different possible governance modes in relation to exerting control (O'Mahony 2007). The overall goal of this dissertation therefore revolves around the question of how firms influence and control OSSDPs they are dedicated to.

## 1.2  Research Questions and Dissertation Goal

Due to its principles that contradict those of proprietary software development, OSS as a phenomenon has attracted increasing attention to researchers and managers in recent years. There have been theoretical and empirical articles published concerning various topics in relation to using an OSS development approach, such as:

- economics of OSS (e.g., Casadesus-Masanell & Ghemawat 2006, Casadesus-Masanell & Llanes 2009, Darmon & Torre 2009, Demil & Lecocq 2006, Economides & Katsamakas 2006, Lerner & Tirole 2002, Perens 2005),

- the adoption and business value of OSS (e.g., Chengalur-Smith, Nevo & Demertzoglou 2010, Chengalur-Smith, Sidorova & Daniel 2010, Rossi Lamastra 2009, Torkar, Minoves & Garrogós 2011, Ven & Verelst 2008, Ven & De Bruyn 2011),

- the relationship between firms and OSS communities (e.g., Bonnacorsi et al. 2006, Capra, Francalanci, Merlo & Rossi Lamastra 2011, Dahlander & Magnusson 2005, Dahlander & Magnusson 2008, Henkel 2009, Krishnamurthy & Tripathi 2009),

- evolving OSS business models (e.g., Hemphill 2006, Krishnamurthy 2003, Krishnamurthy 2005, Mann 2006, Riehle 2011b, Teece 2010b),

- the motivation of individual programmers to provide their labor for free (e.g., Bitzer, Schrettl & Schröder 2007, Hars & Ou 2002, Krishnamurthy 2006, Lakhani & Wolf 2005, Stewart & Gosain 2006, Wu, Gerlach & Young 2007, Xu et al. 2009), and

- organization, structure, and hierarchy within OSS communities (e.g., Cornford, Shaikh & Ciborra 2010, Crowston & Howison 2006, De Laat 2007, Franck & Jungwirth 2003a, Hahn, Moon & Zhang 2008, Iannacci 2005, Markus 2007, O'Mahony & Ferraro 2007, Von Krogh, Spaeth & Lakhani 2003).

Although those articles, and many others alike, did a great job in helping explaining OSS principles and implications for business, they lack explanations of how firms can control the project's trajectory in relation to their interests and investments. For example, regarding the relationship between firms and communities, firms are often viewed as a coherent group with common interests and visions about an OSSDP (Dahlander & Magnusson 2005), ignoring different business models and corresponding interests and control structures those firms may apply.

By studying different motivation structures of individuals providing their service to an OSSDP for free, many researchers added to the understanding of antecedents for voluntary contributions to a public good (e.g., Bitzer et al. 2007, Wu et al. 2007). However, although researchers have discussed the role of firm-sponsored programmers and their motivation structure (e.g., Krishnamurthy 2006, Lakhani & Wolf 2005, Roberts et al. 2006), they largely neglected the existence of different business models and their coordination and control necessities built upon an OSSDP that might influence a firm-sponsored developer's motivation.

Similarly, research that outlines the importance of organizational aspects of OSS development has primarily drawn attention to structures and processes within a community, such as leadership structures, network positions, hierarchy, or core-periphery structures (e.g., Crowston & Howison 2006, Dalander & O'Mahony 2011, Fleming & Waguespack 2007, Giuri, Rullani & Torrisi 2008, Grewal, Lilien & Mallapragada 2006, Lakhani 2006, MacCormack, Baldwin & Rusnak 2010, Masmoudi, den Besten, de Loupy & Dalle 2009). Only a few have taken the presence of an individual's sponsoring by a firm and related control potentials into account.[7] In summary, despite considerable efforts in providing explanations of how OSS development works and which roles firms and their employed developers assume within OSSDPs, recent research primarily has featured an OSS centric view. However, from a firm's perspective, OSS development still is an innovation activity even though development might take place beyond firm boundaries. In this sense, firms have to manage innovation with neither having complete ownership over the product nor

---

[7]See Dahlander & Wallin (2006), Henkel (2009) or Stewart, Ammeter & Maruping (2006) for examples of a few exceptions.

being able to apply labor contract mechanisms for external developers. But how can a firm influence or control an OSSDP it is investing resources in?

Conversely, organizational control theory (e.g., Eisenhardt 1985, Ouchi 1979) features a firm centric view. Furthermore, the focus of organizational control theory is on internal resources such as individuals who are bounded by labor contracts. Resources external to the firm such as a community of developers are beyond the firm's area of direct influence, a situation that is not captured by classical control theory.

Thus, because there is a dearth of research applying a firm centric view to OSS development and because classical control theory (e.g., Barker 1993, Kirsch 1997, Ouchi & Johnson 1978) lacks applicable concepts, a number of questions concerning organizational control are not yet entirely answered. For example, given that firms allocate resources to an OSSDP, such as authorizing employed developers to devote their labor to the project, how are those developers advised to behave within the community? Are organizational structures of a firm mirrored within OSS communities? How does the presence of multiple firms with potentially multiple interests influence a single firm's relation to an OSSDP? And finally, does the business model influence the intensity of firm's engagement in OSS development? Thus, by taking on a firm centric view, all these questions may be subsumed under the overall research question that guides this dissertation:

> If innovation is managed at least partially outside and across the legal and organizational boundaries of the firm, how can a firm influence or even control a project its business model depends on without having discretionary power over developers external to the firm?

This dissertation seeks to answer this question in two steps. First, a theoretical basis will be provided by merging relevant research into OSS, such as the relationship between firm and community, OSS business models, community structure, and knowledge management in open innovation with research into organizational control. Building upon this basis, an extended control theory will be developed that captures the charateristics of innovation activities beyond firm boundaries and in the absence of vertical command chains, such as in the case of OSS. Second, drawing on various notions of authority, the extended control theory for managing innovation activities beyond firm boundaries will be tested in both a multi-project (Chapter 4) and a single-project (Chapter 5) scenario.

As such, this research is, to the best of my knowledge, the first that conceptually defines the different options a firm might apply for controlling OSSDPs they do not own, and, in addition, the first that provides empirical evidence for these options within and across different projects. The contribution of this dissertation is complemented by showing avenues for further theoretical and empirical work. Finally, this dissertation may give recommendations to managers of software vendors planning to adopt the OSS approach of software development.

## 1.3 Anchorage in Philosophy of Science

Before I start outlining the structure of this dissertation and reporting the findings of my investigations, I briefly want to share my understanding of organization and management as well as information systems (IS) research. No one would doubt that one primary goal of a dissertation is to contribute to the creation of scientific knowledge in a specific discipline. However, prior to gaining scientific knowledge through the use of a certain method it seems suitable to rethink norms, values, and procedures that we (as researchers) take for granted.

In front of almost every discussion in or about epistemology is the claim for truth. From a radical constructivist point of view, it is still questionable if there even is an objective truth as our perception of reality is constructed in a brain that is rather isolated and independent from any input from outside (Von Glasersfeld 1995). However, given the assumption that there is a truth, how can we identify and judge what is true and what is not?

With regard to the latter question, philosophers differentiate between different concepts of truth (Frank 2006). *Correspondence theory* of truth treats a proposition as true, if it is consistent with the part of reality it describes. However, as the theory further assumes that a correspondence between a proposition and the analyzed part of the reality can be observed, it is not useful to uncover a superior truth due to the problem of biased perceptions. The *coherence theory* of truth demands testing new knowledge against accepted wisdom and *consensus theory*, which might be viewed as an extension of coherence theory, builds on human judgment by a group of elaborated scientists in order to define if a proposition is true or not. Admittedly, neither of these approaches provides a solution for situations where two different groups of scientists reach contradictory conclusions.

Thus, all abovementioned concepts of truth show deficiencies, especially due to their focus on already existing knowledge. Possibly the most famous example of continuously contributing to an existing body of scientific knowledge ignoring a superior truth is the case of Newtonian mechanics. It was not until Einstein formulated mechanisms of action in cases of speed close to light velocity that fundamentally differed from what people treated

as reality – supported by Newton's observations –, that a whole discipline had to accept
that observation itself is not effectual to unrevealing *the* truth.

Driven by the incompleteness of concepts of truth, generations of philosophers turned
toward investigating the process of generating knowledge. As a result, we find many different approaches to explaining the creation of scientific knowledge. These approaches or
"schools", often derived from different disciplines, are considered guiding paradigms within
science. In order to position the present dissertation in the philosophy of science, I will
briefly sketch the most influential approaches.[8] However, as these approaches are based on
basic ontological and epistemological positions,[9] further clarification is needed first. Burrel
& Morgan (1979) developed a framework based on the distinction between a subjective
and an objective approach to social sciences. In their work, they differentiated between
four different layers, namely ontology, epistemology, human nature, and methodology, mirroring different levels of assumptions. By simultaneously considering different layers and
perspectives, this work portrays the choice of a research method as a function of the initial
choice of perception of reality (see Figure 1.1).

Being the extrema of a continuum, the subjective approach to social science would
consider reality as an individual's mental product whereas the objective approach grasps
reality as external to the individuum. Furthermore, the subjective and objective view on
social science are mutually exclusive perspectives (Von Kortzfleisch 2004), meaning that
they cannot simultaneously operate as a basis for the same research project.

Given the heterogeneity in philosophical approaches to social science, it has increasingly been argued that it would be theoretically unsound if different epistemological and
ontological assumptions are mixed (e.g., Burrel & Morgan 1979, Chen & Hirschheim 2004,
De Vaujany, Lesca, Fomin & Loebbecke 2008, Niehaves 2005). Consequently, if the combination of different epistemological perspectives is restricted by theoretical consideration
within the philosophy of science (Kuhn 1962), following Burrel & Morgan (1979), the
combination of multiple research methods is also restricted, something known as *method
incommensurability*. For example, observation, although deeply anchored in the tradition
of behavioral sciences, would not be applicable in a constructivist view of the world where
a subjective approach to science was chosen.

Referring to influential schools in social science, *positivism* probably is the most widely
adopted school in information systems research (Frank, Schauer & Wigand 2008, Niehaves

---

[8]See Frank (2006) for a detailed discussion on the applicability of different schools of thought to IS research.
[9]Ontology deals with the philosophical study of existence and reality while epistemology describes ways
to get access to reality.

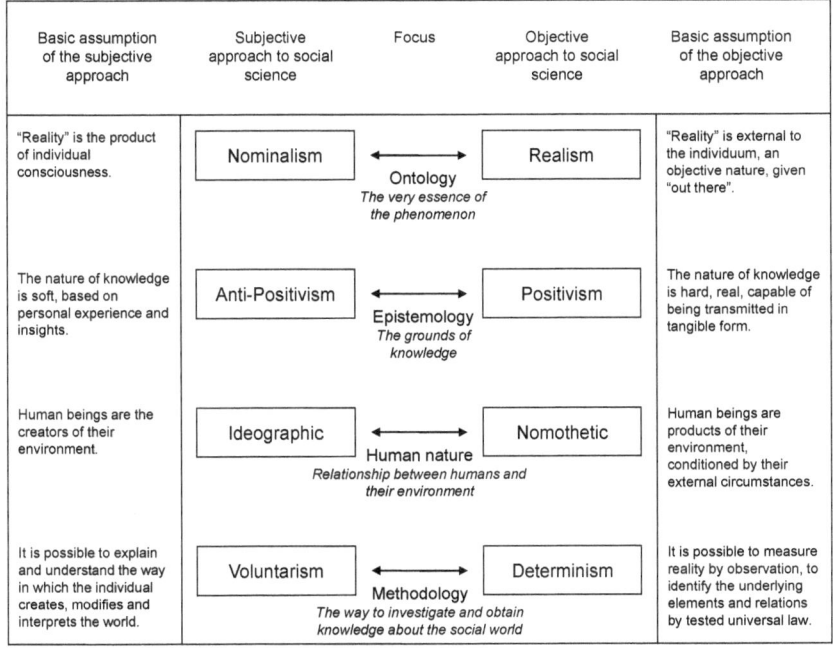

| Basic assumption of the subjective approach | Subjective approach to social science | Focus | Objective approach to social science | Basic assumption of the objective approach |
|---|---|---|---|---|
| "Reality" is the product of individual consciousness. | Nominalism ◄——► Realism<br>Ontology<br>*The very essence of the phenomenon* | | | "Reality" is external to the individuum, an objective nature, given "out there". |
| The nature of knowledge is soft, based on personal experience and insights. | Anti-Positivism ◄——► Positivism<br>Epistemology<br>*The grounds of knowledge* | | | The nature of knowledge is hard, real, capable of being transmitted in tangible form. |
| Human beings are the creators of their environment. | Ideographic ◄——► Nomothetic<br>Human nature<br>*Relationship between humans and their environment* | | | Human beings are products of their environment, conditioned by their external circumstances. |
| It is possible to explain and understand the way in which the individual creates, modifies and interprets the world. | Voluntarism ◄——► Determinism<br>Methodology<br>*The way to investigate and obtain knowledge about the social world* | | | It is possible to measure reality by observation, to identify the underlying elements and relations by tested universal law. |

**Figure 1.1** The Subjective and Objective Dimensions in Philosophy of Science (Source: Burrel & Morgan 1979, p. 9)

2005). In its broadest form, referred to as logical positivism, it is based on the assumption that scientific knowledge only can be based on empirical evidence or proven logic (e.g., Frank 2006). As the word *positive* has the additional connotation of something useful, for instance, reality in contrast to non-reality, positivism therefore is based on realism and an objective approach to reality (Störig 1968). However, in contrast to research in engineering or computer sciences, it is inherently clear that with a positivistic approach to social science, the creation of future worlds is limited, if not impossible, as positivism does not allow for scientific knowledge that is not tested against something already existent.

By denying the assumption of the logical positivism, namely excluding the existence of prior knowledge, Popper (1934) created a stream in philosophy of science known as *critical rationalism*. Depending on how Popper is interpreted, critical rationalism may be viewed as an extension of logical positivism, as it is still based on empiricism, but an empiricism that is grounded on a theoretical background that existed a priori (Settle 1979). Remembering

Humes, Popper further postulated that it is logically impossible to inductively generate theory based on single observations. With regard to positivism in its purest sense, this implies that researchers following a positivistic approach, in theory, would have to perform a complete test against reality. That is, every possible instance of reality has to be tested until results are found that lead to a rejection of the proposed theory.

*Constructivism* considers nature not just as given, but as an aggregation of cultural constructions (Frank 2006). Variations of constructivism, such as radical constructivism, reject a person's ability to objectively perceive reality, as any reality is the product of individual consciousness. As this leads to relatively unfeasible implications for social sciences, alleviated variations, such as the Erlangen constructivism, instead believe in the capability of resolving the problem of constructed reality by developing explicit observation and theory languages (Mir & Watson 2000, Thiel 1984).

Finally, I want to include a philosophical stream in this enumeration of approaches that has received comparatively little attention, namely *constructive empiricism*. A characteristic of this approach is that the acceptance of a theory does not include the blind belief in its universal truth (Van Fraassen 1980, Van Fraassen 2008). Instead, the constructive empiricism only accepts a theory in terms of its empirical adequacy:[10] "Science aims to give us theories which are empirically adequate; and acceptance of a theory involves as belief only that it is empirically adequate" (Van Fraassen 1980, p. 12).

The aim of this dissertation is to develop theory as well as to test this theory against reality. Regarding theory development, this dissertation applies the concept of theory in the sense of Seth & Zinkhan (1991, p. 77) who followed Hunt (1983) and Rudner (1966). They define theory as a:

> systemtiatically related set of statements, including some lawlike generalizations, that is empirically testable. The purpose of theory is to increase scientific understanding through a systematized structure capable of both explaining and predicting phenomena.

Regarding the test of theory against reality, since this dissertation uses empirical data to answer the research questions provided in the precursory section, based on the discussion above, it implicitly adopts a positivistic perspective. However, following positivism in its

---

[10]Based on the distinction between a syntactic and a semantic view of scientific theories, a theory is empirically adequate, if appearances are isomorphic to the empirical substructures of some model of the theory. In other words, the theory is empirically adequate if the observable phenomena can "find a home" within the structures described by the theory. (cf. Stanford Encyclopedia of Philosophy, http://plato.stanford.edu/entries/constructive-empiricism/#1.5, last access: 11/5/2011).

purest sense would imply to deny the acceptance of other contradictory theories (Bird 2003, Spencer 1987). To allow the theory that has to be developed to exist conjointly with other potenially conflicting theories, this dissertation aims to stay coherent with the constructive empiricism provided by Van Fraassen (1980) – although anti-realistic – in that the acceptance of a theory is based on its empirical adequacy only.

## 1.4 Positioning of the Dissertation

The phenomenon of firms engaging in the development of OSS may be viewed from different perspectives, such as economics, management, law, or information systems (Brügge, Harhoff, Picot, Creighton, Fiedler & Henkel 2004). Consequently, this dissertation is influenced by a variety of disciplines, but may be mainly classified as a contribution to organization and management science as well as to information systems research. Each of these disciplines, in turn, consists of a number of different approaches and methods. Whereas information systems research as understood in North America shows no significant difference from management and organization research (Chatterjee 2001), within European information systems research, multiple approaches are applied (Frank et al. 2008). Therefore, in order to ligitimize the choice of the methods used in this dissertation, I will discuss briefly where this dissertation is positioned within information systems research and organization and management science.

Information systems research is claimed to be multidisciplinary as business administration, information science, sociology, and psychology contribute to studying the development, implementation, and usage of information systems and information technology inside organizations (Cecez-Kecmanovic 2011, Chen & Hirschheim 2004, Niehaves 2005, Orlikowski & Baroudi 1991, Wade & Hulland 2004, Winter 2008). Furthermore, as discussed, different schools of thought, especially if embedded in different disciplines, imply the use of different methods. For example, whereas behaviorism, which is grounded on positivism (Danziger 1979), is the dominating paradigm in North American information systems research and turned toward empirical investigations with the aim of describing the nature of reality (Baskerville & Myers 2002, Frank et al. 2008), design science, an alternative approach to information systems research, "creates and evaluates IT artifacts intended to solve identified organizational problems" (Hevner, March, Park & Ram 2004, p. 77).

One of the main differences between behaviorism on the one side and design science on the other, is that the latter generates new knowledge mainly through interpretative logic that results in hermeneutic models or artifacts (March & Smith 1995, Nunamaker, Chen & Purdin 2004, Van Aken 2004). In contrast, driven by empirical results, behaviorists seek to explain reality based on observations (Danziger 1979). However, both paradigms show deficiencies. Whereas design science has difficulties finding theoretical justifications due to its orientation to construction (Baskerville, Lyytinen, Sambamurthy & Straub 2010), behaviorism is blamed for its intense, and sometimes unreflected, use of empirical methods (Baron 2010, McCloskey 1985b, McCloskey & Ziliak 1996, Straub, Boudreau & Gefen 2004) and a "neurotic behavior [...] such as its compulsive handwashing in statistical procedures" (McCloskey 1985a, p. 18).

Consequently, a combination of these approaches therefore would simultaneously benefit from observations represented by empirical results and initiating and creating phenomena that otherwise are difficult to find in reality.[11] Recently, European scholars especially therefore called for an increased use of design-oriented research in information systems in order to complement the process of creating scientific knowledge (Baskerville et al. 2010, Österle et al. 2010). In addition, design-oriented research is increasingly requested even for organization studies (Romme 2003, Von Krogh 2010).

However, although information systems as a discipline, would benefit from methodological pluralism, in order to answer a *single* research question or to close a *single* research gap, using multiple methodologies would be incommensurable and contradict the claim for methodological fit (Edmondson & McManus 2007). Consequently, using multiple methodologies[12] within one dissertation ought to be abandoned. Therefore, aligning the proposed research questions discussed in Section 1.2 with an appropriate methodology is mandatory. As answering the research questions will require observing behavior, in this case behavior of individuals and firms, following a behavioral approach to science is necessary. Thus, this dissertation is based on an understanding of information systems research in a behavioral sense and therefore equally may contribute to management and organization science.

---

[11]It is important to note that while behaviorism is driven by the search for a universal truth within existing realities, design science aims to create realities. However, although design science aims to creation (of artefacts), it not necessarily requires a constructivistic view on reality.

[12]Note: Whereas *methodology* may be interpreted as the theory of methods, *method* refers to systematically giving details of procedures used. Therefore, refaining from using methodological pluralism does not prohibit the use of multiple methods.

# 1.5 Structure and Outline

As seen, **Chapter 1** consists of an introduction to the phenomenon of firms engaging in the development of software that is available for free, a presentation of relevant research questions, thoughts about the importance of philosophy of science for this work, and a positioning of this dissertation in the context of different disciplines. The remainder of this dissertation is segmented into five chapters that I will introduce briefly.

**Chapter 2**, *Managing Innovation Beyond Firm Boundaries*, will deal with different sources of innovation inside and outside the boundaries of the firm and discusses benefits and consequences of opening up innovation processes. In detail, I will provide insights into how innovation emerges and how the value of an innovation is captured and secured (e.g., Pisano & Teece 2007), recent trends in innovation management, such as user and open innovation (e.g., Chesbrough 2003d, Von Hippel 1988) and absorptive capacity (e.g., Lichtenthaler 2009a), and into approaches to align different sources of knowledge with firms' interests.

**Chapter 3**, *Commercializing and Controlling Open Source Software Development*, features the role of a boundary chapter and is devoted to the phenomenon of OSS development and various forms of control beyond firm boundaries. Starting with important principles of OSS and accompanying characteristics, this chapter contributes to a coherent understanding of OSS, shows why OSS is different from other innovation approaches, and provides implications for firms and volunteers. This includes aspects such as motivation of voluntary and professional developers (e.g., Lakahni & Wolf 2005, Shah 2006, Wu et al. 2007), advantages and disadvantages of using and contributing to OSSDPs (e.g., Alexy 2009), licenses and license choice (e.g., Chen et al. 2007), and finally, open source business models (e.g., Perr, Appleyard & Sullivan 2010, Riehle 2011b, Wesselius 2008).

Additionally, the chapter puts emphasis on the importance of organization theory in order to understand appropriate management approaches used for governing OSS projects. Therefore, it encompasses discussions of organizational control (e.g., Cardinal 2001, Kirsch, Ko & Haney 2010, Ouchi 1979, Ouchi & Johnson 1978, Scholl 1999), organization and network structures to be found in distributed innovation teams (e.g., Crowston 1997, Crowston & Howison 2006, Long & Siau 2008, Piccoli & Ives 2003), and different tensions that emerge through the presence of firms in OSSDPs, such as intended collaboration versus increased need for control (e.g., Almirall & Casadesus-Masanell 2010, Sundaramurthy & Lewis 2003), or such as the simultaneous use of management by design (within the boundaries of the firm) and management by community (within the community) (e.g., Amin & Cohendet 2004). The chapter will close with remarks that classical management and governance approaches are only of limited use for distributed innovation modes such as OSS development (Dahlander & O'Mahony 2011). Finally, drawing upon these remarks, an extended control theory for innovation activities beyond firm boundaries and in the absence of vertical command chains will be developed.

Chapters 4 and 5 may be viewed as a second part of the thesis. Here, I will develop hypotheses about behavior of firms and individuals within OSSDPs. In **Chapter 4**, which is called *Open Source in Action I: Business Collaboration Among Open Source Projects*, I will investigate the use of different governance modes using the example of Eclipse, an OSS foundation that support the commercialization of OSSDPs. By considering different types of projects, namely those who are dominated by a single firm and those who receive contributions by multiple firms, I show how the OSS approach can be used as an alternative to research and development (R&D) alliances or joint ventures. In addition, the empirical investigation aims to unveiling different control modes firms apply according to the type of project.

Complementing the work presented in chapter 4, **Chapter 5**, *Open Source in Action II: Business Collaboration Within an Open Source Project* draws attention to participation of voluntary and professional OSS programmers within the development of a single OSSDP. Using social network theory (e.g., Freeman 1979, Sparrowe, Liden, Wayne & Kraimer 2001, Tushman & Scanlan 1981b, Wasserman & Faust 1994), this second investigation contributes to the discussion on open source innovation by considering the influence of an individual's network position within a community on the number of his technical contributions. The analysis is based on publicly available Email-data and data from source code files of the Linux kernel developer community. Linux, an OSS operating system, exists in many derivatives, but the kernel is maintained by a designated group of developers including the founder of Linux, Linus Torvalds (Dafermos 2001, Moon & Sproull 2002). The analysis shows that an individual's performance is dependent on his network embeddedness that provides him with horizontal authority and whether or not he is sponsored by a firm. The chapter ends with a discussion of the results in light of existing theory and related work.

Finally, **Chapter 6** summarizes my contributions, provides implications for theory and management, and sketches avenues for further research.

# Chapter 2

# Managing Innovation Beyond Firm Boundaries

## 2.1 Defining the R&D Boundaries of the Firm

As discussed in the introductory chapter, this dissertation aims to explore the role of firms and their employed developers within OSS communities. I further argued that the business model a firm is applying in order to benefit from OSS adoption, development, or distribution, is equally important to understand the phenomenon. However, since software development in general may be viewed as a form of innovative activity (Buxmann, Diefenbach & Hess 2008), being able to classify possible results of this dissertation against existing knowledge on innovation management requires a full understanding of how innovation emerges within and outside the research and development (R&D) boundaries of the firm. As current management thinking is predominantly influenced by the resource-based view of the firm (RBV; e.g., Barney 1991, Makadok 2001, Peteraf & Barney 2003) and new institutional economics, which is an economic perspective on social and legal norms that determine the choice of integrating external resources or not (e.g., Frese 2000, Samuels 1998, Valcárcel 2002), I will start with a discussion about firm boundaries in the light of new institutional economics, RBV, and appropriability regimes.

### 2.1.1 New Institutional Thoughts on Firm Boundaries

New institutional economics has its origins in the theory of the firm as formulated by Coase (1937). His insights, drawn from thoughtful comparisons with price mechanisms as coordination instruments, deliver evidence for the existence of firms and further highlight the interplay of the number of transactions and firm size. The term new institutional economics, however, was created by Williamson (1975), who contributed seminal work by considering contracts in the context of transaction costs. Today, new institutional economics is considered as an umbrella for various (economic) theories of the firm (e.g., Kieser & Walgenbach 2010, Scott 2008, Suddaby 2010), such as

- transaction cost theory,

- principal-agent-theory, and

- property rights theory.

**Transaction Cost Theory.** In economics, a transaction is understood as any exchange of property rights or resources (Milgrom & Roberts 1992). For example, if person A wants to sell a good and person B is willing to give an adequate monetary compensation for that good, a transaction takes place. Williamson (1975) argued that any transaction produces costs on both sides, the demand side and the vendor side. He further distinguished between *ex-ante* transaction costs, such as search, information, and bargaining costs, and *ex-post* transaction costs, such as control costs. In turn, transaction cost theory puts emphasis on the contract as an organizational form in order to appraise the formation of institutions as a function of frequency, specificity, uncertainty, limited rationality, and opportunistic behavior (Williamson 1996). In other words, *transaction cost theory* revolves around the question which kind of institution, understood as an organizational form, minimize the total cost (including transaction and production costs) of producing and distributing a good or service under conditions of outcome uncertainty and opportunism of agents. For example, repeatedly agreeing on contracts that entail uncertainty increases the risk for cost-intensive renegotiations and thus provides an incentive for vertical integration (Pisano 1990).

The choice of a contractual framework, therefore, is determined by transaction costs rather than by price mechanisms. Make or buy as well as keep or sell decisions, which were primarily discussed in the light of the price for a good or service, may now be discussed based on appropriate governance mechanisms for certain transactions (Lichtenthaler 2010b, Silverman, Nickerson & Freeman 1997). Thus, markets and hierarchies are discrete structural alternatives for any transaction and each is supported by a different contractual arrangement (Demil & Lecocq 2006, Powell 1990, Williamson 1991).[1] Whereas hierarchy

---

[1] Powell (1990) discussed "network" as a third alternative to market and hierarchies, which I will turn to in Chapter 3.

is based on employment contracts that allow for governance mechanisms like authority (Aghion & Tirole 1997), markets are supported by classic contract law, by which control is difficult to employ until courts are appealed (Demil & Lecocq 2006).

**Principal-Agent Theory.** The *principal-agent-dilemma* is a direct consequence of information asymmetry between one entity (principal) that considers hiring another entity (agent) (e.g., Eisenhardt 1989, Green & Stokey 1983, Holmstrom 1979). In principle, only the agent has complete information about what he has done in order to complete a requested task. Until the outcome is consistent with what the principal expected, the agent is likely to minimize his effort to reach predefined goals as the risk to bear consequences is limited. In other words, as long as the principal is not able to perfectly observe the agent's behavior, the agent is encouraged to choose risky actions, such as extending breaks, he would not have undertaken had he been confronted with the full risk. This behavior is also known as *moral hazard*, that is, a change of behavior in the absence of perceived risk (Holmström 1979). In addition, the agent might show opportunistic behavior, such as distorting important information. In summary, due to information asymmetry, it is almost impossible for the principal to judge the level of an agent's degree of goal achievement. Admittedly, the principal can overcome this deficiency by installing monitoring and control levels (Boatright 2004). However, installing different control layers and monitoring systems is costly, and a principal would withdraw from hiring an agent if monitoring costs exceeded the benefits resulting from employing the agent (Jensen & Meckling 1976).

By considering multiple agents to be hired by a principal, as in the case of joint production where agents are employed and bounded by contract, Alchian and Demsetz (1972) argued that it is difficult to determine each worker's contribution and to pay accordingly. If the payment for each worker is equal and dependent on the total output, each worker has an incentive to minimize his effort to contribute to goal achievement (Boatright 2004). As existing literature on incentives and contract design is concentrated on a single principal-agent-relationship, that is, a relationship between one principal and one agent, Green & Stockey (1983) discussed tournaments as an alternative to contracts in cases of multiple agents. Instead of rewarding agents based on absolute performance, they proposed a rewarding system based on relative performance. The agent, or employee, respectively, then has an incentive to do better than his peers. The principal-agent-theory therefore has a clear link to transaction cost theory as it is determined by a contract-based view (Foss 1996).

**Property Rights Theory.** According to Demsetz (1967, p. 347), property rights entail the privilege to benefit or harm from harnessing the rights. For example, "harming a competitor by producing superior products may be permitted, while shooting him down may not. A man may be permitted to benefit himself by shooting an intruder but be prohibited from selling below a price floor." Thus, property rights regulate who may be benefited or harmed (Henry 1999). Following a resource-based definition, "property rights to resource attributes consist of the rights to use, consume, obtain income from, and alienate these attributes" (Foss & Foss 2005, p. 542). Attributes, in this sense, consist of different functionalities and services assets can supply (Penrose 1959). For example, an owner of a car is permitted to change the speed of his car, but this right may be constrained within pedestrian zones or in areas with speed limitations. The owner of a good therefore holds property rights over the good's attributes, but his rights may be constrained by external contractual arrangements.

Regarding ownership, it is important to distinguish between *de jure* and *de facto* property rights (Bromley 1991, Demsetz 1988). Theoretically, one can use, consume, and obtain income from attributes of a good or resource one does not own. For example, if someone rents a flat, he receives the property right to use it, but will not become the owner. Thus, an economic actor may obtain property rights over goods or resources without having ownership. However, in order to fully benefit from having property rights, according to Alston and colleagues (2009), a full set of property rights is required, which includes:

1. the right to use the asset in any manner that the user wishes, generally with the caveat that such use does not interfere with someone else's property right,

2. the right to exclude others from the use of the asset,

3. the right to derive income from the asset,

4. the right to sell the asset, and

5. the right to bequeath the asset to someone of your choice.

As a full set of property rights never exists because it would be too costly to enforce the right against unauthorized usage (Coase 1960), some attributes of goods or resources remain open, which may be obtained *de facto* or *de jure* by third parties (Barzel 1989). Furthermore, especially in cases where ownership structures are undefined, such as the atmosphere or high seas, property rights are distributed, resulting in problems like overuse of the asset, known as "tragedy of the commons" (Alston, Harris & Mueller 2009).

In addition, property rights are means to internalize external effects and externalities (e.g., Bromley 1991, Henry 1999, Liebowitz & Margolis 1994). Here, externality is understood as any consequence of economic decisions nobody pays for or receives a compensation for. Those effects may be either pecuniary and determine the distribution of income, such as losing paying customers due to increased competition, or non-pecuniary, such as environmental damage. Internalization of externalities thus implies transferring property rights from one economic actor to another. As a consequence, benefit-costs relationships change. Therefore, economic actors prefer to internalize externalities only if the gains of internalization exceed the costs of internalization (Demsetz 1967).

As demonstrated, new institutional economics attempts to explain the emergence of firm boundaries by considering contractual frameworks in which transactions take place. From a new institutional perspective, organizational forms, such as market and hierarchies, are not the result of price mechanisms, but the result of efforts in minimizing transaction costs within given contractual frameworks. Furthermore, stressing the importance of transactions envisions that the exchange of property rights is more important to an economy than the exchange of goods per se (Coase 1988, Foss & Foss 2005, Picot, Reichwald & Wigand 1998). As the principles of new institutional economics may be applied for any transaction within and between firms and with regard to any department, such as human resources, finance, or procurement, using a new institutional approach in order to understand the genesis of R&D boundaries of a firm suggests itself.

## 2.1.2 A Resource-Based View on R&D Boundaries of the Firm

Besides being influenced by new institutional economics, current management thinking is affected by another dominating approach, namely the resource-based view of the firm (RBV). In contrast to opposing approaches that use privileged market position to explain a firm's competitiveness (e.g., Porter 1980), the RBV posits that those firms that possess or have access to a collection of superior tangible and intangible resources gain competitive advantage (e.g., Barney 1991, Dierickx & Cool 1989, Verona 1999, Wernerfelt 1984). Here, the term resource is not limited to the economic understanding of resources as factors of production, such as land, labor, and capital goods (e.g., Gutenberg 1983, Sullivan & Sheffrin 2003). Moreover, resources are stocks of assets that are owned or controlled by the firm and may be defined as any virtual or physical entity of limited availability that needs to be consumed to obtain benefit from it (Amit & Schoemaker 1993).

Consequently, the RBV attempts to explain that performance differences result from variation in resources and capabilities across firms (e.g., Barney 2001, Easterby-Smith & Prieto 2008, Levinthal & Wu 2010, Mata, Fuerst & Barney 1995, Mol & Wijnberg 2011). Aspects of the RBV to which research has given attention in the past include internal resources, such as skilled and knowledgeable labor (e.g., Nahapiet & Ghoshal 1998, Nonaka & Takeushi 1995, Prahalad & Hamel 1990) or external resources, such as suppliers, partnerships, information technology, networks, or employees of other firms (e.g., Colombo, Rabbiosi, & Reichstein 2010, Mathews 2003, Schaarschmidt, Von Kortzfleisch, Valcárcel & Lindermann 2011).

Thus, being able to develop, deploy, and protect unique combinations of competences and resources is a source of competitive advantage. In a similar vein, Teece, Pisano & Shuen (1997, p. 510), who developed the concept of "dynamic capabilities" argued that

> because this approach [dynamic capabilities] emphasizes the development of management capabilities, and difficult-to-imitate combinations of organizational, functional and technological skills, it integrates and draws upon research in such areas as the management of R&D, product and process development, technology transfer, intellectual property, manufacturing, human resources, and organizational learning.

Regarding internal resources, various researchers put emphasis on the importance of employees with their labor and knowledge for firm success (e.g., Hayes, Wheelwright & Clark 1988, Nonaka & Takeushi 1995). In a similar vein, in his 1962 book *Strategy and Structure*, Chandler pointed to the importance of labor and stated that "trained personnel with manufacturing, marketing, engineering, scientific, and managerial skills often become even more valuable than warehouses, plants, [or] offices..." (Chandler 1962, p. 383). In addition, start-ups or small and medium sized enterprises (SME), which typically lack complementing resources such as financial capabilities, rely especially on skilled and knowledgeable workers as well as on employees' capability to go the "extra mile" in order to gain competitive advantage (Bessant & Tidd 2007, Mosakowski 1998, Van de Vrande, De Jong, Vanhaverbeke & De Rochemont 2009, Von Kortzfleisch & Mergel 2002). Investors seeking to invest in start-ups, therefore, highlight the composition of entrepreneurial teams as a major criterion for investment decisions (Stevenson & Jarillo 1990, Von Kortzfleisch, Schaarschmidt & Magin 2010, Yehezkel & Lerner 2009).

Admittedly, deploying resource bundles on an employee level that consist of difficult-to-imitate combinations of organizational, functional, and technical skills requires advanced training and education of existing personnel as well as the acquisition of external labor. As a consequence of investing in human resources by sending a firm's own employees to training courses or hiring highly educated workers, a firm simultaneously can extend its knowledge base. In turn, as a firm's knowledge base may be considered a resource, a firm, especially firms active in R&D intensive industries, such as information technology, biotechnology, or aviation industry, may gain competitive advantage by focusing on maintaining their knowledge bases (Hansen, Mors & Løvås 2005, Singh & Soltani 2010, Zhang, Shu, Jiang & Malter 2010). As a result of emphasizing knowledge as a unique resource, a knowledge based-theory of the firm (KBV) emerged within the larger framework of the RBV (Grant 1996b).

However, not only are current and future employees "carriers" of knowledge, but skilled and knowledgeable workers are valuable sources of ideas that potentially may be turned into economic value (Kijkuit & Van den Ende 2007, Van Dijk & Van den Ende 2002). Thus, as the ability to develop innovative ideas is a critical success factor in rapidly changing markets (e.g., Harhoff & Mayrhofer 2010, Poole, Van de Ven, Dooley & Holmes 2000) and dependent on available human resources, firms that want to strengthen their market position are well advised to focus on the composition of their human resource pool (Kogut & Zander 1992).

Consequently, if unique bundles of resources determine a firm's competitiveness, firms are likely to integrate as many available resources such as skilled and knowledgeable labor as possible – without considering transaction costs economics. Thus, unlike new institutional economics that focus on contractual frameworks, the RBV explains the emergence of the (R&D) boundaries of the firm as a function of acquiring valuable resources and releasing unwanted ones. By putting emphasis on resources, managerial implications drawn from considering transactions retreat into the background (Foss & Foss 2005) and the role of a firm's core competencies is invigorated (Prahalad & Hamel 1990). Similarly, in one of his later publications, Chandler (1992, p. 86), although he largely avoided using the term RBV, formulated that "the nature of the firm's facilities and skills becomes the most significant factor in determining what will be done in the firm and what by the market."

Like other theories of the firm, the RBV has its limitations. For example, implicit in the RBV is that a market for resources such as a market for ideas, knowledge, technology or labor exists *per se* (e.g., Arora 1995, Arora & Gambardella 2010, Dushnitsky & Klueter 2011). If such markets would not exist, exchange of resources would not be possible. However, by referring to the non-existence of a market for corporate reputation, which is considered an important resource of a firm (Bolton, Ockenfels & Ebeling 2011, Walsh

& Beatty 2007, Weigelt & Camerer 1988), Dierickx & Cool (1989) remarked that not all resources may be bought and sold.[2]

In addition, although the RBV is valuable for explaining differences among firms' performances, it gives relatively little attention to the cost side.[3] Building, maintaining, and acquiring strategically relevant resources is still costly. For example, developing a valuable corporate reputation requires investments in advertising and customer relationship programs; headhunting valuable employees of other firms requires offering intensive, such as higher salaries, and acquiring the permission to use technological knowledge, for instance, in the form of licenses for patent use (e.g., Fosfuri 2006, Gambardella & Hall 2006, Lichtenthaler 2011c) is equally cost incentive. Therefore, combinations of resources do not create monetary value *per se*; instead, they only create monetary value for a firm when benefits exceed the costs incurred for building them (Wassmer & Dussauge 2011).

To sum up, the RBV may be considered an alternative explanation for the emergence of firm boundaries and differs from other theories of the firm. Firstly, unlike new institutional economics, the RBV aims to expose reasons for differences in firms' performance. Secondly, by focusing on bundles of resources and capabilities rather than transactions and contractual arrangements, the RBV regards decisions that affect the composition of firm boundaries as a result of concentrating of core competences. In that sense, outsourcing (e.g., Barthelémy 2011) is not (only) a question of transaction costs, but a question of whether the outsourced resource is strategically important to a firm or not. Since the RBV, as well as the new institutional perspective, both have their advantages and disadvantages for explaining a firm's competitiveness and the emergence of firm boundaries, recent publications aim to combine aspects of both theories (e.g., Foss & Foss 2005).

### 2.1.3   Appropriation and Appropriability Beyond Firm Boundaries

In their ambition to combine aspects of the RBV and new institutional economics, Foss & Foss (2005, p. 543) suggested that it would be "useful to think of resources as bundles of property rights. [...] How property rights are constrained by the law, agreements, and norms, influence how much value a resource owner can create and appropriate from the

---

[2]Dierickx & Cool (1989) further argued, that due to market inefficiencies, strategically relevant resources in general only may be accumulated internally. However, recently emerged markets for technology licensing and innovation intermediaries demonstrate the opposite.

[3]For example, counting for the word "cost" in Barney (2001) delivered exactly *one* result, in Makadok (2001) *none*.

resource." As this quote puts emphasis on the value of resources in relation to property rights, it acts as an adequate start for a discussion on value appropriation and protection. Firms that have invested in building, maintaining and acquiring resources tend to protect their investments in various forms (Abernathy & Utterback 1978, Dushnitsky & Lenox 2006, Lukach, Kort & Plasmans 2007). To put it simply, a firm has to ensure that other parties are excluded from using the property rights of resources' attributes the firm owns. However, different types of resources require different protection mechanisms as some resources entail natural protection while others do not. For example, knowledge about the processes needed to produce wine is more difficult to imitate than knowledge on how to run a saloon. Regarding IP, which may be viewed as one of the most important resources in a knowledge based economy (Grant 1996a, Hagedoorn, Cloodt & Van Kranenburg 2005, Teece 2000), protection appears through intellectual property rights (IPRs). Patents (i.e., a protection mechanism of knowledge embedded in novel ideas), copyrights (i.e., a protection mechanism of knowledge embedded in art and literature), and trademarks, (i.e., a protection mechanism of knowledge embedded in symbols), all are means of legally protecting the use of IP (Andersen & Konzelmann 2008, Mansfield 1986, Schilling 2008).

Against this background, those firms gain competitive advantages that manage to limit competitors' ability to appropriate from innovations enabled by the focal firm. However, to fully benefit from innovation, it is not sufficient just to establish restrictions on the external use of resources, such as IP (Lewin 2007). A firm additionally has to find ways to appropriate itself from the value it has created by utilizing its resources. In his seminal work, Teece (1986, p. 287) therefore pointed to an innovator's[4] ability to capture the profits generated by an innovation as a function of his embeddedness in *regimes of appropriability*. An appropriability regime is considered *weak*, if the technology[5] is almost impossible to protect and *tight* (or *strong*) if it is relatively easy to protect. Since in weak appropriability regimes competition tends to increase as the innovation is easy to imitate, it has widely been argued that a strong appropriability regime may act as an incentive for knowledge creation (e.g., Dahlander 2005, Helpman 1993, Shavell & Van Ypersele 2001, Sojer 2011). In the absence of strong appropriability regimes, however, firms that seek to appropriate from innovation rents have to rely primarily on speed, time to market, and luck (Katz &

---

[4]For Teece (1986), an innovator is not an individual, but a firm that has first commercialized a new product.

[5]Although Teece (1986) used the term "technology" here, from an IP perspective, it is the knowledge embedded in a technology rather than the technology itself that is about to be protected.

Shapiro 1986, López & Roberts 2002). For example, Groupon, a start-up company that offers coupons to customers that have subscribed to a newsletter, builds on a business model without IPR protection as the idea of selling vouchers is highly imitable. Groupon responded to the pressure of increased competition with a speedy growth that made it, according to *Forbes Magazine*, the fastes growing company ever.[6] In a similar vein, companies that have built their business model around the sale of services miss strong appropriability regimes (Edvardsson, Tronvoll & Gruber 2011, López & Roberts 2002, Teece 2010b).

Although IPRs are powerful means to secure the innovator's ability to capture rents from innovation, they are sometimes either not applicable (e.g., in the case of services) or provide ineffective protection (e.g., in case a patent only covers parts of a technology). Thus, the weakness or tightness of the appropriability regime is *not* a direct consequence of the IP protection mechanisms (Ghosh 2005). In turn, firms that do not have ownership or control over IPRs may still be able to capture a share of the profits gained by the innovation. For example, if a follower manages to identify and use gaps in the IPR system, as in the case of many inventions[7] produced and protected by Xerox (e.g., mouse driver, network card), he will receive rents based on a technology he neither invented nor first commercialized (Chesbrough 2003b). Moreover, late entrants with competitive advantages in terms of pre-existing market shares, access to financial resources, or marketing and reputation advantages may outperform the inventor if the IPR system is inefficient. Coca Cola and Pepsi, for instance, were not the first to introduce cola in a can, but succeeded in the long run because they took advantage of their dominating market position. In summary, IPR itself is not efficient to secure a strong appropriability regime.

In his search for a theory of value appropriation, Teece (1986) further emphasized the role of complementary assets in the process of commercializing inventions. Based on the distinction between tacit and codified knowledge about a technology, Teece argued that this knowledge somehow must be sold in a market in order to benefit from the invention sufficiently. However, although technological knowledge on a stand-alone basis may be sold to other firms[8] (e.g., Arora 1995, Lichtenthaler 2007, Lichtenthaler & Ernst 2007), in

---

[6]Forbes Magazine: Meet The Fastest Growing Company Ever, http://www.forbes.com/forbes/2010/0830/entrepreneurs-groupon-facebook-twitter-next-web-phenom.html, last access: 11/30/2011

[7]The distinction between invention and innovation goes back to Joseph Schumpeter (1936). Whereas *invention* refers to the creation and establishment of something new, *innovation* is used to describe inventions that became economically successful (Erwin & Krakauer 2004).

[8]Some firms specialize in developing new ideas and technologies and avoid bringing their ideas to the market themselves. For instance, Abgenix developed break-through inventions in the biotechnology field, but avoided developing, manufacturing, and marketing possible products (Schilling 2008, p. 153f).

general, firms that want to commercialize the technology themselves (for instance, in the case of consumer goods) will require additional capabilities and assets, such as marketing, manufacturing, and after-sales services (Rothaermel & Hill 2005).

In his work, Teece (1986) further differentiated between generic, specialized, and cospecialized complementary assets. Whereas generic assets are applicable to multiple contexts (e.g., a production line that is able to handle multiple products and innovation), specialized and cospecialized assets are connected to a unique innovation. Generic complementary assets thus share certain characteristics of platforms. Cospecialized assets are characterized by bilateral dependencies while specialized assets are characterized by unilateral dependencies, meaning that the asset is either dependent on the innovation or the innovation is dependent on the asset. For example, software is usually dependent on specific hardware. Adobe's flash player, for instance, is dependent on a hardware infrastructure, which is not provided by Apple. For Adobe, any hardware delivered by Apple thus is a specialized complementary asset to which it does not obtain access. Apple, on the other side, is not dependent on Adobe's flash player as alternatives exist, such as HTML5[9]. In a situation with balanced dependencies, Teece used the example of container ships and ports; from the innovation's perspective, complementary assets are cospecialized (Tripsas 1997). Figure 2.1 provides an illustration of the concept of complementary assets.

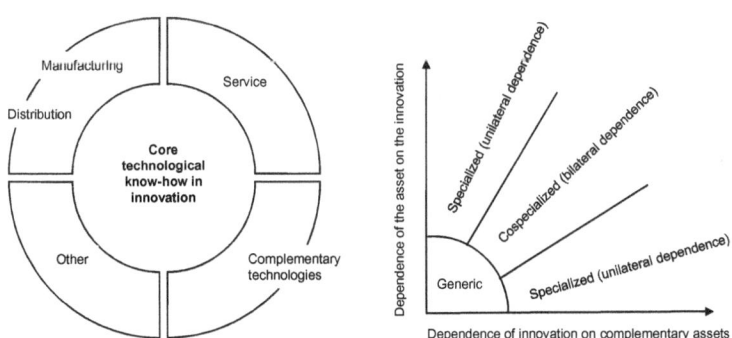

**Figure 2.1** Generic, Specialized, and Cospecialized Complementary Assets (Source: Teece 1986, p. 289)

---

[9]It is worth noting that Adobe, in turn, is not dependent on Apple either. There are other providers of hardware platforms that accept the flash player standard. Apple denies the application of the flash player to its platforms due to strategic reasons.

Complementary assets, as understood by Teece, are more than complementing products required to fulfill a customer's need (e.g., pasta and tomato sauce). Since almost any capability or asset that is of importance for appropriating rents is considered a complementary asset, the concept acts as a framework for explaining innovation success. Firms that own, control, or have access to the entire bundle of complementary assets gain competitive advantage by minimizing competitors' ability to capture value shares of the innovation. It is important to bear in mind, that it is the control over strategically important assets and not the IPR that protect the innovation.

However, owning and controlling the entire bundle of complementary assets is costly, especially when firms are active in different networks that require different combinations of them (Bekkers, Duysters & Verspagen 2002). Thus, important complementary assets or capabilities will be partly, or even completely, under external control. Despite the protection against imitation, firms have to align with external stakeholders that may influence or even be a part of the appropriability regime. In some situations, limiting protection efforts may even be advantageous as the speed of diffusion tends to increase when many parties promote an innovation (Abrahamson 1991, Fichman 2000, Tuppuraa, Hurmelinna-Laukkanena, Puumalainena & Jantunena 2010).

To sum up, whereas IPRs are means to protect innovation against competitors, the concept of complementary assets helps to reveal why firms fail even though they have established strong IPR systems. It is important to bear in mind, that the locus of this discussion is the innovating firm. By shifting the focus to the imitator, it becomes obvious that protection is only necessary if there is someone with an interest to appropriate from that innovation. Therefore, whereas the innovator's aim is to protect its innovation and to own or control complementary assets, the imitator is likely to enforce getting access IPRs and assets that complement an innovation. In line with the RBV, if complementary assets are considered a necessary resource, the emergence of the boundaries of the firm may be explained by internalizing resources relevant to capture value from innovation while externalizing irrelevant ones.

However, as recent evolutions have shown, complementary assets, capabilities, or resources do not have to be internalized necessarily. For example, firms active in the development of OSS do not employ all developers even though this would give them perfect control, both from an RBV and a transaction cost perspective. In a similar vein, Dahlander (2005, p. 282) posited that "these firms face the challenge that important resources for finding competitive products and services are located outside the boundary of the firm and in the public domain." Thus, as owning, controlling, or having access to crucial assets, capabilities, and resources has different implications for the composition of the innovating firm (e.g., the structure of the R&D department), drawing the boundaries of the firm becomes increasingly difficult. In addition, recent phenomena, such as user generated and open innovation (e.g., Chesbrough, Vanhaverbeke & West 2006, Von Hippel 2005) contributed to the permeability of firm boundaries, which raises defining the R&D boundaries of the firm to one of the most challenging tasks in modern management.

## 2.2 Opening Firm Boundaries for Innovation

Knowledge, which is needed to complement existing knowledge in order to innovate, is often not readily available in the organizational knowledge base. Thus, firms are explicitly searching for knowledge and innovative ideas from outside the corporate boundaries. By opening their innovation processes to external sources, firms enlarge their internal knowledge base and obtain additional competitive advantage (e.g., Awazu & Desouza 2004, Bogers 2011). These external sources, which can be customers, suppliers, universities and even competitors, have long been excluded from the firm's innovation process (Chesbrough 2003d). Usually, sources that contribute to the generation of value are not available for free, and firms that intend to use such sources pay accordingly. However, in recent years, due in large parts to technological accomplishments such as the Internet, users have provided feedback on technologies or come up with own ideas – at no cost to the firm (Dahlander & Frederiksen 2011). Moreover, in some industries (e.g., windsurfing or outdoor sports) users even innovate, that is, they not only formulate ideas but they produce physicl components themselves and bring them to markets (e.g., Baldwin, Hienerth & Von Hippel 2006, O'Hern & Rindfleisch 2010, Sawhney, Verona & Prandelli 2005).

On the other hand, in some situations the firm produces knowledge spillovers, that is, the firm develops more knowledge than it can transform to innovations (O'Mahony & Vecchi 2009, Panagopoulos 2009). In these cases, firms still may gain revenues if they are able to license unused knowledge out (Lichtenthaler 2009b). Open innovation, not yet an organization theory, is a paradigm that aims to combine both views; complementing missing knowledge through external resources on the one hand and managing to profit from unused knowledge on the other.

## 2.2.1    Customer Co-creation, User, and Community Innovation

Building on the distinction between invention and innovation, an invention occurs if existing knowledge is recombined or is enhanced by complementing knowledge. The result, though, is a new product or procedure that has to be accepted by a market in order to be named innovation (Schumpeter 1936). The process of recombining knowledge is still perceived as mysterious and in their attempt to shed light on different antecedents, various researchers have pointed to the relevance of creativity for innovation (e.g., Martins & Terblanche 2003, Tassoul 2006). Creativity, defined as "the production of novel and useful ideas in any domain" (Amabile, Conti, Coon, Lazenby & Herron 1996, p. 1155), thus, is an important step in the process of transforming available knowledge into successful innovations (Van Dijk & Van den Ende 2002). Research about creativity has mainly focused on employees in a traditional R&D environment who are embedded in a hierarchical organizational structure. For example, it has been shown that intrinsic motivation is positively related to employee creativity (e.g., Amabile 1985, Amabile & Gryskiewicz 1989, Dewett 2007). However, creativity, as well as complementing knowledge may be located outside the firm boundaries (Carlile 2004, Coff 2003).

The availability of creativity and knowledge outside the boundaries of the firm, thus, challenges the predominant view of the creation of innovation within closed R&D departments (Lichtenthaler 2011b). However, as stressed by Trott & Hartmann (2009), innovation hardly arises in complete isolation. The demand-pull perspective of market creation and emergence, for instance, argues that customers define their needs before firms put effort into developing an adequate solution to fulfill that need (White 1981). As supporters of the demand-pull explanation claim that a market will be created primarily when possible customers are asking for the fulfillment of specific needs (Chidamber & Kon 1994), this view underlines the importance of the customer's voice as an external input to internal R&D activities.

Von Hippel (1978b) was one of the first who highlighted the shift of customers from passive consumers to active co-creators. His customer active paradigm challenged the predominant marketing and market research view at that time that was based on the principle that customers only "speak when spoken to" (Von Hippel 1978a, p. 243). Building on the seminal work of von Hippel (1988), various researchers from different disciplines contributed to the discussion of user participation and involvement – shifting the focus to individual customers as innovators. For instance, Dahan & Hauser (2002) distinguish three ways of how to manage customer interaction and cooperation, namely listening, asking, and taking part, with taking part being the most open method. Nambisan (2002) analyzed different customer roles as related to a specific phase in a new product development (NPD) process, with customer as resource in the "ideation"-phase, customer as co-creator in the "design and development"-phase and customer as a user in the "testing and support"-phase. Thus, with the emergence of the Internet, and especially with the phenomenon of user-generated content, the role of a customer, or user, respectively, in the innovation process had to be reconsidered as a co-creator of value in various scenarios (Edvardsson et al. 2011, Füller 2006, Harrison & Waluszewski 2008, Nambisan & Nambisan 2009, Neyer, Bullinger & Moeslein 2009, Oliveira & Von Hippel 2011, West & Bogers 2010).

Today, firms actively involve or even integrate users in their innovation process in several ways (Kleinaltenkamp 2002). For example, a firm may use customers' knowledge without any effort on their part (e.g., reading customers' blogs). In another scenario, a firm may actively promote contributions of possible current and future customers, such as in the case of idea contests (e.g., Hutter, Hautz, Füller, Mueller & Matzler 2011), *lead user* integration (e.g., Lüthje & Herstatt 2004, Olson & Bakke 2001, Von Hippel 1986) or mass customization (e.g., Reichwald & Piller 2006). Consequently, various researchers have pointed to the potential of virtual customer environments or toolkits, that is, firms outsource need-related innovation tasks to the user by providing them access to user friendly design tools that they can use to develop parts of the innovation themselves (e.g., Nambisan & Baron 2009, Piller & Walcher 2006, Thomke & Von Hippel 2002, Von Hippel & Katz 2002).

In some industries, such as the skateboarding or windsurfing industry, users even were the creators of the entire innovation, including the commercialization of their product (Shah 2000). Long before commercial firms became aware of the potential of these innovations, users themselves became entrepreneurs and therefore have been the creators of those markets (Shah & Tripsas 2007). As individual customers are able to take on different roles in the innovation process, such as, according to Nambisan (2002), customer as resource, customer as co-creator, and customer as user (innovator), the boundaries between users and producers have dissolved (Franke & Shah 2003, Lüthje & Herstatt 2004, Rothwell 1986).

In order to distinguish users from producers, Baldwin & von Hippel (2009, p. 3) argued that "users, as we define them, are firms or individual customers that expect to benefit from using a design, a product or a service. In contrast, producers expect to benefit from selling a design, product, or service." A more detailed distinction is provided by West & Bogers (2010). Their work, which may be considered an extension of Nambisan's (2002) conceptualization, even though they did not refer to it, is based on the differentiation between an *input*, *self*, *share*, and a *startup* mode of user innovation. The *user innovation (UI)-input* mode assumes that user innovations act as an input for new or improved commercial products.[10] *UI-self* refers to situations where users innovate to solve their own needs – without reporting their findings to a firm or sharing them with a community of other users (Von Hippel 2007). In principle, any change to a product already in a user's possession would fall into this category. Sharing a user's own innovations without regard to commercialization prospects is defined as *UI-share* by West & Bogers (2010). In *UI-share*, users share their innovations both *horizontally* with the community independently from any sponsoring firm or *vertically* with upstream producers (De Jong & Von Hippel 2009). Finally, *UI-startup* refers to users that have become entrepreneurs themselves, meaning that they not only created the innovation, but commercialized it as well. Despite their rather unstructurerd handling of the term innovation (e.g., if they would follow Schumpeter's perspective, any user innovation would fall into the UI-startup mode by definition), West & Bogers (2010) drew a very precise picture of different aspects of user innovation.

As highlighted by West & Bogers (2010) and many other researchers (e.g., Bogers 2011, Franke & Shah 2003, Harhoff, Henkel & Von Hippel 2003, Sleeswijk-Visser, Van der Lugt & Stappers 2007), sharing innovation, or at least what is considered to be innovative, is an important part of user innovation. Users rarely innovate in complete isolation, as they are – like innovating firms – dependent on complementary assets. Depending on the user innovation mode, these assets may be peers who promote the idea (e.g., in UI-share) or managerial and marketing capabilities (e.g., in UI-startup). In other scenarios, a community of users collectively develops an idea, or innovation, respectively. Based on the seminal work of Allen (1983) on *collective invention*, various researchers therefore have made community innovation the focus of their study (e.g., Harhoff & Mayrhofer 2010, Hutter et al. 2011, O'Mahony 2005, Osterloh & Rota 2007, Van de Ven & Garud 1993). For example, Dahlander & Frederiksen (2011) were able to show, that a user's innovativeness (rated by experienced judges in conjunction with the user community) is dependent on his position within the community as well as on his activities in other user communities. Thus, in contrast to (pure) user innovation, interaction with peers not only enriches the perspectives and ideas of an individual, but combined efforts may result in even better solutions, such as in the case of OSS.

---

[10]UI-input therefore is what other streams of research call *user involvement* (e.g., Barki & Hartwick 1994, Ives & Olson 1984) or *customer integration* (e.g., Kleinaltenkamp 2002).

To sum up, users, which are usually simultaneously customers of a firm, assume different roles in an innovation process. From the firm's point of view, they either operate as an input for firm-driven innovation processes (e.g., by providing needs or ideas) or may even be an integral part of the innovation process (e.g., by co-designing or co-developing). Thus, depending on the degree of integration, users supplement important activities usually performed by the firm. However, task partitioning and a modular innovation process are crucial to "outsource" parts of the innovation process to users (Von Hippel 1990). Except for the rather seldomly pure form of user innovation (UI-startup),[11] where an established firm is not able to benefit directly, users, or customers, respectively, therefore complement, and in a few cases even supplant the integrated process of innovation commercialization (West & Bogers 2010).

As discussed, firms increasingly make use of external sources in order to supplement or even complement their own innovation efforts. So far, Section 2.2.1 has focused on individual customers and users organized in communities as external sources of innovation and as co-creators of value. However, user involvement in firm-driven NPD is predominant in early and late phases and firms only make limited use of user integration in core development phases (Gassmann, Kausch & Enkel 2010, Schaarschmidt & Kilian 2011). Despite the potential inherent in integrating users' knowledge in own innovation processes, a majority of users simply does not have the required experience and technological knowledge in order to contribute to the development of a product sufficiently. Thus, whereas the integration of individual customers may be of value in Business-to-Consumer (B2C) markets and in early and late phases of the NPD (e.g., input of ideas, testing final product), especially in Business-to-Business (B2B) markets and in the development phase (cf. Cooper 1994), firms rather rely on interfirm collaboration, that is, firms organize themselves in R&D alliances or interfirm networks (Schilling & Phelps 2007).

## 2.2.2 Knowledge Spillover and External Technology Exploitation

Although collaboration on the firm level is cost-intensive due to high transaction costs (e.g., search for partners, negotiations), in the case of vertical integration of external resources, the value of obtaining access to technological knowledge exceeds by far the economization of outsourcing parts of the innovation process to users. In other words, for core research and development activities, firms obviously prefer to pay for getting access to external

---

[11]Despite the relevance of user innovation in modern innovation management, only a few examples exist other than in outdoor and sports industries.

resources, such as existing technologies or other firms' knowledge, over integrating individual, comparably inexperienced users at zero costs (e.g., in the case a user freely provides an idea to the firm). Decades of research therefore has concentrated on investigating the emergence, retention, and decomposition of strategic alliances (e.g., De Rond & Bouchikhi 2004, Sampson 2005, Vanhaverbeke, Duysters & Noorderhaven 2002).

More recently, organization theorists especially have put emphasis on technology and knowledge transfer between organizations (e.g., Brusoni, Prencipe & Pavitt 2001, Lambe & Spekman 1997, Sakakibara 2002, Simonin 2004). For example, by looking at firms' patent portfolios, Mowery, Oxley & Silverman (1996) examined if overlapping technological resources are a result of prior alliance participation. From a knowledge perspective, firms are not only collections of tangible and intangible resources[12] that develop over time, but collections of knowledge workers. In a similar vein, another stream of research highlights the importance of learning – both on the individual and the organizational level – in order to be able to profit from knowledge exchange. Based on the pioneering work of March (1991), various researchers adopted *exploration* and *exploitation* as learning strategies (e.g., Raisch & Birkinshaw 2008, Sidhu, Commandeur & Volbera 2007). Whereas *exploration* refers to the search for new and useful adaptations, *exploitation* refers to the use and propagation of known adaptations (He & Wong 2004). However, due to limited available resources it is unlikely that a firm searches for new ideas and leverages existing adaptations simultaneously to the same extent. In theory, firms either focus on exploitation, which supplies the firm with required returns, but minimizes the number and quality of new adaptations, or on exploration, accepting that searching for the next solution intensely will slow down its current performance. As exploitation usually yields immediate returns, firms generally tend to overemphasize the use and propagation of existing adaptations (Fang, Lee & Schilling 2010). Thus, balancing exploration and exploitation appears to be an important task in technology management. In order to resolve the dilemma, recent research points to organization design as a means to achieve the required balance (Belderbos, Faems, Leten & Van Looy 2010, Ethiraj & Levinthal 2004, Siggelkow & Rivkin 2006).

Although the majority of studies considers exploration as an activity that can be pursued inside the boundaries of the firm (e.g., building their own R&D departments) and outside (e.g., external technology acquisition), admittedly, an underlying assumption is that firms "leverage their knowledge by means of internally developing products and ser-

---

[12]As knowledge can be implicit or explicit (Nonaka & Takeushi 1995), a firm's resources embrace tangible and intangible aspects of knowledge.

vices" (Lichtenthaler & Ernst 2009b, p. 371). However, in some situations, as a result of intense search for new and useful adaptations, the firm may have generated more knowledge than it is able to commercialize. For example, in the 1980s, Xerox invented breakthrough technologies in its R&D lab PARC[13], such as a mouse driver or a precursor of Microsoft's Word, but was unable to bring it to market because they focused on promoting their core business, i.e. selling printers (Chesbrough 2003b, Chesbrough & Rosenbloom 2002).

Chesbrough refered to the overproduction of technological knowledge as "knowledge spillover." Knowledge spillovers, in turn, may be considered the result of overemphasizing exploration[14] and knowledge acquisition (Audretsch & Feldman 1996, Jaffe 1986). On the other hand, knowledge spillovers are a necessity for external knowledge commercialization.[15] In a similar vein, West & Gallagher (2006, p. 319) pointed to the option of externally commercializing unused technological knowledge but simultaneously advise against missing IPR protection:

> For while some IP that could not be internally commercialized was licensed to others, all too frequently it 'sat on a shelf' waiting either for internal development, its research proponents to leave the firm to develop it on their own, or even more dangerously, for it to 'spillover' to other firms.

However, whereas spillovers to a direct competitor may be disadvantageous in the same industry, spillovers to firms that help in creating a new market may be economically rational in some situations (Brandenburger & Nalebuff 1996, Panagopoulos 2009).

Exploration, as well as exploitation may be viewed from two different perspectives, an internal and an external one. As discussed above, the internal view posits that both, exploration and exploitation occurs within the boundaries of the firm, such as when the firm commercializes its internally developed innovations. Regarding the external view, consequently, exploration revolves around activities such as knowledge acquisition, whereas exploitation refers to external technology commercialization, such as out-licensing. However, compared to the intensity of studies on external knowledge exploration, research on external knowledge exploitation has been neglected (Yli-Renko, Autio & Sapienza 2001). In addition, Lichtenthaler (2005), who was one of the first who systematically analyzed external knowledge commercialization, stated that literature on external knowledge exploitation

---

[13]PARC stands for Palo Alto Research Center
[14]It is worth noting that Chesbrough largely avoided the term "exploration" in his work.
[15]I want to thank Henry Chesbrough for clarifying this point and for underpinning the relevance of knowledge spillovers for open innovation during the "PhD Seminar on Open Innovation" at ESADE Business School, Barcelona, January 2011.

is rather fragmented. For example, "knowledge", "technological knowledge" and "technology" are confusingly used interchangeably. However, although technological knowledge is embedded in technology and sometimes even protected by IPR (e.g., patents), knowledge refers to a broader applicability. Lichtenthaler himself partly used "technology marketing" as a synonym for "external knowledge exploitation", but later précised his understanding: "We use the term 'external knowledge commercialization (exploitation)' instead of 'knowledge (technology) marketing', because it allows for the integration of external knowledge exploitation into a more comprehensive approach to knowledge management" (Lichtenthaler 2005, p. 232).

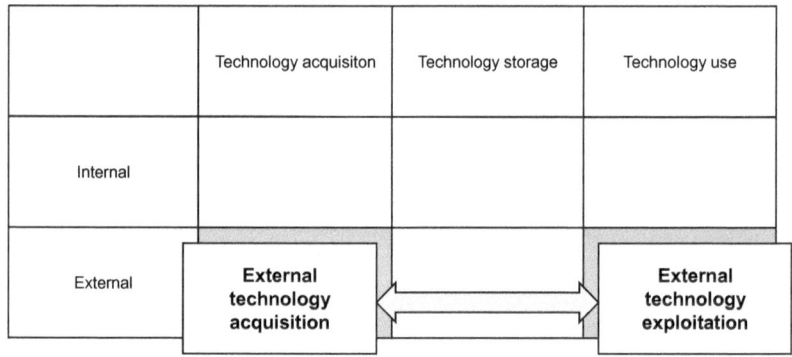

**Figure 2.2** External Technology Acquisition and Exploitation (Source: adapted from Lichtenthaler 2005, p. 233)

Lichtenthaler, who complemented and enriched the literature on external knowledge (technology) exploitation (e.g., Lichtenthaler 2007, Lichtenthaler 2009b, Lichtenthaler 2010b, Lichtenthaler 2010c), further provided an integrated framework of different management tasks in the context of internal and external exploration and exploitation (see Figure 2.2). Based on the distinction between internal and external, technology acquisition is distinguished from technology storage and use. Lichtenthaler placed greater concern on the use of "technology" as the level of analysis and prefered "knowledge" in most of his frameworks (cf. Lichtenthaler 2005, Lichtenthaler 2011c, Lichtenthaler & Ernst 2006), but with regard to the topic of this dissertation, namely the role of firms in the technical development of OSS, the use of the term "technology" is favored here.

However, it is worth noting that despite the recently increasing penetration of the concept in academia, firms only make limited use of the external technology exploitation option. For example, Lichtenthaler (2010a) reported a mean value 2.40 on a 7-point Likert-type scale for the extent of external technology commercialization in his investigation of 154 medium-sized and large industrial firms in Germany, Austria, and Switzerland.[16]

In summary, external technology acquisition, such as buying patents for their own use (cf. Jones, Lanctot & Teegen 2001) and external technology exploitation, such as licensing out unused patents and IP (cf. Lichtenthaler, Ernst & Conley 2011) are thus means of managing innovation that is either developed or commercialized beyond firm boundaries. Knowledge spillovers, although considered to be dangerous if deficiently protected, are a prerequisite for externally commercializing knowledge or exploiting technologies. By drawing on the dilution of firm boundaries in two directions (i.e., inside-out and outside-in), the Lichtenthaler framework not only puts emphasis on an integrated approach to knowledge/technology exploration and exploitation as learning strategies, but additionally may be considered an important foundation to fully understand firms' opening decisions.

### 2.2.3 Open Innovation as an Integrative Perspective on Firm Boundaries

In the past, firms mainly followed the predominant industrial logic of closed innovation strategies with the focus on internal aspects like the development of technological knowledge to be applied in their own products or services (Chesbrough 2003d, Lichtenthaler 2009a, Lichtenthaler 2010c). In order to secure the ability of internal knowledge creation, firms consequently invested in their own resources or hired the smartest people available. In recent years, firms have started to rethink these strategies and begun to rely on sources from inside and outside the boundaries of the firm. This trend is referred to as *open innovation* meaning that in contrast to a closed innovation paradigm, where firms pursue developing, markteting, financing, and supporting the product on their own (see Figure 2.3), firms try to include customers, users, universities, and even competitors in different stages of their new product development processes (Laursen & Salter 2006). Thus, in those companies, the paradigm of a more internal and closed R&D organization is dissolved in favor of a common and collaborative process between companies and its external stakeholders.

In order to delineate open from closed innovation, some important steps in innovation management history are briefly reported. According to Chesbrough (2003c, p. 37), who introduced the term, "Open innovation is a paradigm that assumes that firms can

---

[16]Note: The aim of Lichtenthaler's investigation was not to show the limited use of external technology exploitation, but to show it connectedness to external technology acquisition.

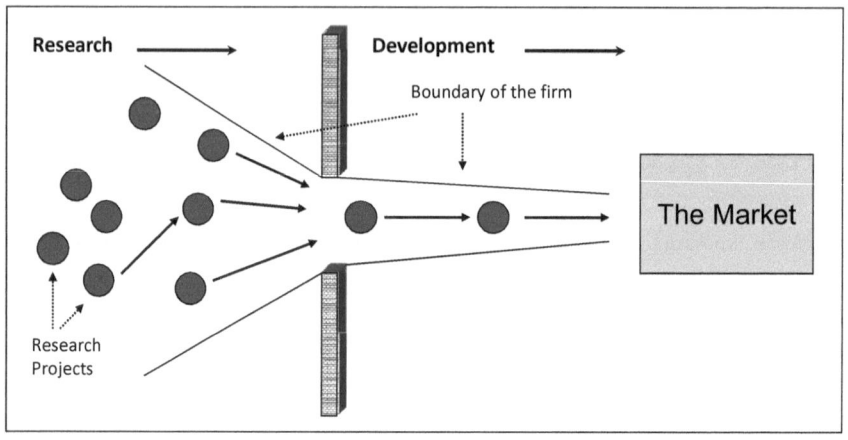

**Figure 2.3** The Closed Innovation Model (Source: Chesbrough 2003c, p. 35)


and should use external ideas as well as internal ideas, and internal and external paths
to market [...]." Later, Chesbrough expanded his definition of open innovation by highlighting the importance of knowledge: "Open innovation is the use of purposive inflows
and outflows of knowledge to accelerate internal innovation, and expand the markets for
external use of innovation, respectively" (Chesbrough et al. 2006, p. 2). However, due
to an overemphasis on knowledge inflows, partly because open innovation was intensely
related to the user innovation phenomenon – partly because frequently cited publications,
such as Laursen & Salter (2006), have concentrated on the breadth and depth of search activities – the knowledge outflow side was somewhat neglected. At first glance, this neglect
is reminiscent of the overemphasis of exploitation over exploration as discussed in Section
2.2.2. Admittedly, whereas the overemphasis of exploitation is a result of a firm's decision
for immediate returns, in contrast, the overemphasis of knowledge inflows is a result of
different research focuses in academia.

Although the number of research articles on open innovation has increased rapidly in
recent years as shown by Dahlander and Gann (2010), the basic principles are new neither
to management research nor practice (Trott & Hartmann 2009) but can only be understood
by combining different management theories, like the resource-based view, organizational
learning theory, or transaction cost economics (Vanhaverbeke, Cloodt & van de Vrande
2008). As revealed in numerous studies and research articles, which often combine different

perspectives, external sources like users or strategic partners have always been a valuable input for numerous innovation processes and have resulted in a variety of innovations (e.g., Allen 1983, Ahuja & Katila 2001, Chesbrough 2003a, Fey & Birkinshaw 2005, Kogut & Zander 1992, Teece 1986, Von Hippel 1994). In a similar vein, the less intensely investigated external commercialization of internally developed intellectual property has also always been the focus of management research (e.g., Lichtenthaler & Ernst 2007, Shane & Ulrich 2004, Von Hippel & Von Krogh 2003).

However, although a major research stream still focuses on either the integration of external sources or the exploitation of internally developed products or services, building on Chesbrough's work, various researchers began to subsume both views under the term open innovation (e.g., Chiaroni, Chiesa & Frattini 2009, Chiaroni et al. 2010, Elmquist, Fredberg & Ollilia 2009, Laursen & Salter 2006, Lichtenthaler 2009b, Keupp & Gassmann 2009, Saur-Amaral & Amaral 2010, Van de Vrande, de Jong, Vanhaverbeke & de Rochemont 2009, Van de Vrande, Vanhaverbeke & Gassmann 2010). The remarkable thing about open innovation, therefore, is that researchers simultaneously devote attention to an outside-in, and an inside-out perspective. Figure 2.4 provides an overview on different possible knowledge flows across the boundaries of the firm, which, if managed correctly, may even result in new markets.

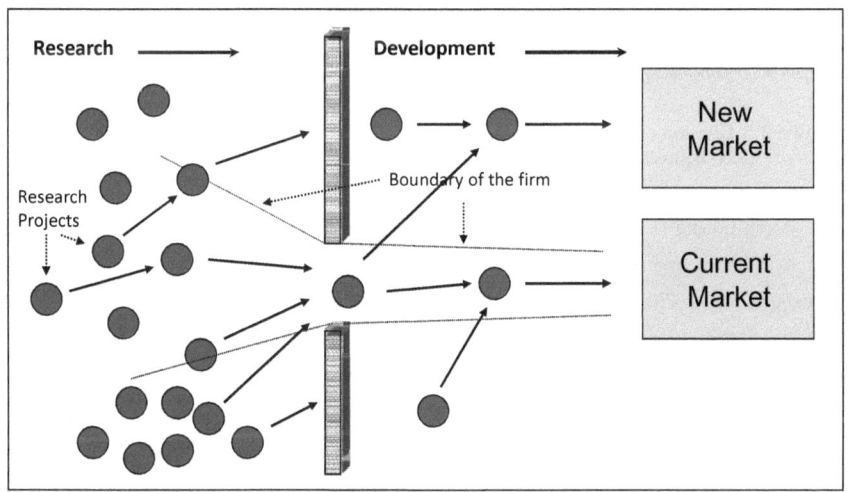

**Figure 2.4** The Open Innovation Model (Source: Chesbrough 2003c, p. 37)

Only a few studies have treated open innovation as more than just a phenomenon and put emphasis on its role as a broader framework to analyze firm behavior. Regarding the different options offered by the open innovation paradigm, Enkel, Gassmann, and Chesbrough (2009) differentiated between three types of open innovation processes, namely (1) outside-in, (2) inside-out and (3) coupled, meaning that although the locus of innovation is still located inside the boundaries of the company, in the case of (1), external knowledge helps to foster innovation and in the case of (2), technology will be exploited outside the company if the developed knowledge does not fit the innovation strategy. Large enterprises are especially likely to combine both archetypes of innovation processes (3) (see Figure 2.5).

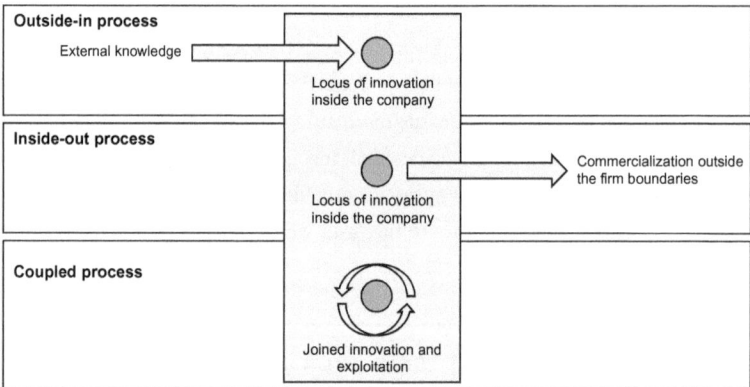

**Figure 2.5** Division of the Locus of Innovation from Knowledge Creation and Commercialization (Source: Gassmann & Enkel 2004, p. 6)

It is worth noting that firms have inherently made use of these processes ever since, but that it is the degree of openness that separates closed from open innovation approaches. Laursen & Salter (2006) as well as Keupp & Gassmann (2009) further empirically showed that different archetypes of open innovation firms exist regarding the breadth and depth of open innovation activities. Open innovation therefore has to be understood as a continuum and not as a binary variable with open and closed as the only possible states. In addition, as openness is viewed as a necessary complement to internal R&D, one of the major challenges for understanding open innovation is to balance between internal and external sources as well as between internal and external paths to markets.

Recently published work aims to develop a more complete view on open innovation from a company's perspective. Rohrbeck, Hölzle & Gemünden (2009), for example, were able to draw a precise picture of open innovation activities at Deutsche Telekom. Although still exploratory in nature, their findings help to understand how different innovation processes are combined along the corporate's value chain and how they vary in terms of intensity.

Dahlander & Gann (2010) provided another open innovation model, which has the potential to act as a framework for further research. Based on an intense literature review, they identified four major open innovation approaches, which can be understood as an extension of Enkel et al.'s (2009) work. In one dimension, they differentiated between outbound and inbound innovation, meaning that outbound refers to how internal resources are revealed to the outside world and inbound refers to how firms can use external sources of innovation. In a second dimension, they distinguish pecuniary from non-pecuniary goals in order to emphasize that not every transaction is related to monetary exchange. The combination of these dimensions leads to types of openness that Dahlander & Gann (2010) termed *revealing, selling, sourcing,* and *acquiring* (see Table 2.1).

**Table 2.1** Structure of Four Different Forms of Openness (Source: Dahlander & Gann 2010, p. 702)

|              | Inbound innovation | Outbound innovation |
| ------------ | ------------------ | ------------------- |
| Pecuniary    | Acquiring          | Selling             |
| Non-pecuniary | Sourcing          | Revealing           |

In summary, open innovation is a paradigm that devotes its attention to knowledge flows in and out of the firm. Thus, it may be considered a framework for any interaction with stakeholders in any direction. Compared to work on explorative and exploitative learning, research on open innovation is rather phenomenon-driven. Recently, researchers have begun to combine research on the open innovation phenomenon and seminal theoretical perspectives, such as dynamic capabilities (e.g., Teece 2009, 2010a), but such approaches are seldom. Therefore, from a theoretical point of view, one of the provoking challenges in future research will be to extent the open innovation phenomenon to a serious theory of the firm and to show, how well established concepts (e.g., explorative and exploitative learning, interfirm alliances) can be integrated in such a meta-theory.

This is important because if opening the boundaries would always be beneficial to a firm, the boundaries of the firm not only would delute but should disappear entirely.

However, despite Chesbrough's claim for opening firm boundaries to accelerate innovation, we still observe that firms are skeptical about lowering their boundaries too much. The reasons for not opening an innovation process vary. Recent publications highlight the importance of IP in the decision-making process of opening firm boundaries, arguing that firms with a competitive advantage through superior knowledge tend to rely on closed innovation processes (e.g., Lichtenthaler 2010a). In addition, firms often do not have the ability to absorb external knowledge because they either fail to identify the right knowledge or, in case they have already identified a valuable external source, they are unable to integrate it in their innovation process (Cohen & Levinthal 1990, Huang & Rice 2009, Zahra & George 2002). The latter often is a consequence of the "not-invented-here" (NIH) syndrome, which occurs when overemphasis on internal knowledge is not inhibited (Lichtenthaler & Ernst 2006).

As research with the aim to integrate various theoretical streams in the open innovation paradigm is still in its infancy, open innovation is only of limited use to explain performance differences or the emergence of firm boundaries. Thus, although we observe an increased opening of firm boundaries, open innovation as a concept is still insufficient to explain firm behavior. In addition, open innovation only provides limited managerial recommendations about how to deal with increased openness exactly, which I will discuss in the following section.

## 2.3   Organizing for Innovation Across Firm Boundaries

As argued by various researchers, firms increasingly open their boundaries to the outside world in order to stimulate their innovative performance and to increase their revenues (e.g., Elmquist et al. 2009). Simultaneously, those firms are captured in a trade-off as opening firm boundaries calls for new management techniques in response to newly emerged challenges, such as the changing role of IP (Chesbrough 2003c).

Despite the considerable attention researchers have given to the open innovation phenomenon,[17] the paradigm itself provides only a few recommendations for how a firm should be organized internally when confronted with the benefits and drawbacks of openness. For example, Chesbrough himself pointed to the fact that high-technology companies "do little

---

[17]Several special issues in well-known journals such as *R&D Management* (in 2009) and *International Journal of Technology Management* (in 2010) have been devoted to the topic.

to share their business model with their researchers, and usually locate their R&D personnel away from the people who plan and execute the strategy" (Chesbrough 2003c, p. 44). However, as stressed by Vanhaverbeke, Cloodt & Van de Vrande (2008), established theoretical work, such as absorptive capacity (e.g., Cohen & Levinthal 1990) or dynamic capabilities (e.g., Teece et al. 1997), if linked to open innovation, may help explain how a firm has to be positioned in an open environment.

The aim of this section, therefore, is not only to highlight the challenges a firm is confronted with by opening its boundaries, but to merge several theoretical streams to arrive at an integrative perspective on openness of firm boundaries and the organizational capabilities required to manage that openness.

### 2.3.1 Challenges of Managing Innovation Externally

In his pioneering work, Chesbrough depicted several challenges that stem from changing conditions in established markets and from how development and commercialization of innovations is pursued. These changing conditions, which Chesbrough named "erosion factors", encompass the growing mobility of highly educated and skilled personnel or the growing presence of venture capital (VC) investment (Chesbrough 2003c). For example, in a closed modus, a firm that has invested in its own R&D activities sought breakthrough technologies that secured the firm's revenue stream. If the firm successfully commercializes that technology, parts of the profit is reinvested for the next breakthrough technology. In this scenario, scientists, whose inventions are not commercialized by the firm, have only limited options to commercialize the technology themselves due to the necessity of high investments in peripheral infrastructure (e.g., manufacturing, marketing, etc.). As a result, those scientists stayed (frustrated) in the company and the uncommercialized technology got dusty stored in an internal shelf.

With increased VC activity, which is observable even for OSS (Schaarschmidt & Von Kortzfleisch 2010), scientists have more options for externally commercializing the invention, to which they might have devoted themselves for a long time. However, as successful start-ups usually achieve initial public offering (IPO) or might be acquired by other firms (Dushnitsky & Lenox 2006, Gruber & Henkel 2006, Hellmann 2006, Hellmann & Puri 2002), the firm that originally funded the invention will not benefit from its commercialization.[18]

---

[18]These arguments describe, based on Chesbrough's (2003c) observations, a pre-open innovation period. Today, technology vendors would neither give venture capitalists easy access to their own inventions, nor would they let scientists and engineers go without signing non-disclosure agreements.

Unintended knowledge outflow as discussed above shows how firms are challenged by changing environmental conditions. Thus, according to Chesbrough (2003d), open innovation is the firm's natural response to increased knowledge transactions, namely, to proactively open firm boundaries instead of passively accepting their dilution. By focusing on knowledge in- and outflows, open innovation is challenging established ways of how knowledge is managed internally. Admittedly, by looking at knowledge flows in particular, several questions arise. For example, given the semiotic view of knowledge, which distinguishes knowledge from information and data or the distinction between implicit (non-codified) and explicit (codified) knowledge (Aamodt & Nygard 1995, Davenport & Prusak 1998, Li & Gao 2003), it is unclear if data, information, and knowledge require the same format of absorption. Whereas knowledge is protectable by IPRs, data might only be protected on a technical layer, not a legal one.

In addition, the firm usually is not a monolith, but a collection of different units and departments. IBM, for instance, keeps Watson research centers in Yorktown Hights, NY, Hawthorne, NY, and Cambridge, MA, as well as other research centers in Brazil, China, Israel, Japan, and Switzerland. In firms with internationally distributed R&D laboratories, internal knowledge management is difficult per se as sometimes different units work on similar problems without exchanging their results. For example, O'Mahony, Cela Diaz & Mamas (2005) described the creation of the Eclipse development platform as a direct consequence of consolidation efforts within IBM. Regarding external knowledge, identifying *where* knowledge flows in and out of the firm is even more difficult in cases of distributed R&D (Hagedoorn et al. 2005, Lichtenthaler & Lichtenthaler 2004, Simonin 2004). Furthermore, as knowledge flows occur on the individual, project, and organization level (Lichtenthaler 2011c, Siggelkow & Rivkin 2006, Siggelkow & Rivkin 2009), firms need additional management capabilities for each mode.

Thus, traditional approaches to knowledge management are challenged by openness. Regarding the value of that openness, two assumptions are underlying the concept that might lead to erroneous interpretations:

1. External knowledge is always superior.

2. IP always has an inherent value.

Chesbrough (2003c) emphasized that the fact that firms increasingly open their boundaries neither implies that external knowledge is superior in any case nor that it substitutes internal knowledge. Thus, firms in an open innovation modus are not only confronted with the challenge of balancing exploration and exploitation (see Section 2.2.2) but are additionally confronted with the challenge of balancing internal and external search strategies (Rivkin & Siggelkow 2003).

Considering the second assumption, knowledge creation is indeed cost and resource intensive. Thus, resulting IP should possess economic value and firms should be likely to protect their knowledge with patents, for instance. However, although there is latent economic value in IP or patents, respectively, without a business model that transforms that latent value into revenue, IP is worth precisely nothing. Thus, the assumption that technology has an inherent value is simply incorrect (cf. Chesbrough 2003c). This implies that if a technology is not commercialized, this is because it does not fit the current business model. Firms in that situation would have to change their business model, which might be a dangerous path (Chesbrough 2006). For example, Xerox has been criticized for not commercializing its valuable technologies, such as the networking card. However, as Xerox's chief technology officer Mark Myers said (cf. Chesbrough 2002, p. 834):

> As a $20 billion company Xerox needs to add $2-3 billion in new revenues every year. [...] the typical payout of $10-$50 million from a successful spinout wouldn't 'make a dent' and [...] we just can't meet our growth goals using these spinout mechanisms.

Thus, firms that devote themselves to the open innovation paradigm are faced with a number of challenges. Firstly, the firm has to find the "right" external knowledge that complements the existing knowledge base and can contribute to the firm's success. Secondly, once the required external knowledge is identified, it has to be internally accepted. The non-acceptance of technology that was not developed inside the R&D boundaries of the firm is known as the NIH-syndrome and refers to the phenomenon that scientists and engineers are often skeptical toward the use of external technological inventions in their own products (Lichtenthaler & Ernst 2006). Thirdly, in order to employ the technology, its commercialization has to be aligned with the current business model. Alternatively, the firm might innovate the business model itself (Chesbrough 2010). Finally, the firm has to manage the external use of technology that does not align with the current business model, for instance, by out-licensing the technology.

In summary, although open innovation is an adequate response to several factors that erode traditional businesses, opening the boundaries of the firm leads to new challenges. As these challenges affect different stages of the NPD, different organizational layers and different aspects of knowledge in- and outflows, an integrative perspective is needed to identify management capabilities required to operate in an open innovation era.

## 2.3.2　A Firm's Absorptive and Knowledge Management Capacity

Regarding the capabilities needed to handle external knowledge properly, various researchers have pointed to the importance of a firm's absorptive capacity (e.g., Cohen & Levinthal 1990, Flatten, Engelen, Zahra & Brettel 2011, Rothaermel & Alexandre 2009). According to the seminal work of Cohen & Levinthal (1990, p. 128), absorptive capacity should be defined as "the ability of a firm to recognize the value of new, external information, assimilate it, and apply it to commercial ends." In addition, they not got tired reemphasizing that this capability is a function of a firm's prior knowledge. After its initiation in the early 1990s, the concept was widely promulgated and was used to explain organizational phenomena on multiple levels, such as organizational learning, industrial economics, and dynamic capabilities (Zahra & George 2002).

However, despite the prevalence of the construct in both theoretical and empirical work (Lane, Koka & Pathak (2006) reported that more than 900 peer-reviewed articles have used the concept), absorptive capacity is a rather surprisingly intangible construct mainly because the initial definition was comparatively broad. Consequently, since broad definitions open the door for wide-stretched interpretations, the concept was interpreted differently. For example, Cohen & Levinthal (1990) spoke of prior related *knowledge* as a prerequisite for absorbing external *information*. Successive work, though, debated absorptive capacity as an ability to absorb external *knowledge* – instead of *information* (e.g., Huang & Rice 2009). As a consequence, in the context of a firm's absorptive capacity, information and knowledge are used interchangeably, which limits the comparability of implications.

In a similar vein, different approaches to operationalize absorptive capacity were applied. One approach considered absorptive capacity as a function of the extent of prior knowledge within the firm, especially research on technology focused on R&D-related operationalizations, such as R&D intensity or size of patent base (Lane et al. 2006). Although, for instance, a huge patent base may indicate an elaborated internal knowledge base, in turn, a firm's internal knowledge base is not necessarily reflected by patents. Consequently, firms that sell large amounts of unused patents would be claimed to have only limited prior related knowledge, even though their market knowledge is an indispensable prerequisite to successfully commercialize their bedded and unused technological knowledge. More recently, researchers have therefore begun to adopt perspectives of learning and dynamic capabilities to cover the entire range of the concept. Lichtenthaler (2009a, p. 822), although still focusing on industrial firms, contributed to the extension of the concept by identifying "technological and market knowledge as critical components of prior knowledge in organizational learning processes of absorptive capacity."

In order to provide a delineation of absorptive capacity for this dissertation I adopt the understanding of Zahra & George (2002, p. 186) in that a firm's absorptive capacity is defined as "a set of organizational routines and processes by which firms acquire, assimilate, transform, and exploit knowledge to produce a dynamic organizational capability." Each of the four tasks, namely *acquiring*, *assimilating*, *transforming* and *exploiting* embrace a set of components. For example, according to Zahra & George (2002), *acquisition* of external knowledge is not only a function of prior related knowledge (which enables a firm to identify external knowledge) but additionally a function of prior investments, as well as of intensity, speed, and direction of effort expended. *Assimilation* refers to the process of internalizing external knowledge. This includes actions such as, analyzing, interpreting, and understanding incoming external knowledge. *Transformation* is a learning related stage in this concept and is considered to be the process of combining the knowledge a firm already possesses with the assimilated external knowledge. Finally, *exploitation*, as already discussed in section 2.2.2, encompass all capabilities needed to leverage technological knowledge.

By applying a knowledge perspective, Zahra & George (2002) distinguished potential from realized absorptive capacity. In particular, potential absorptive capacity is influenced by external sources that, once acquired and assimilated, may complement a firm's existing knowledge base. However, it is not until transformation and exploitation have taken place that the potential absorptive capacity has been realized (Figure 2.6).

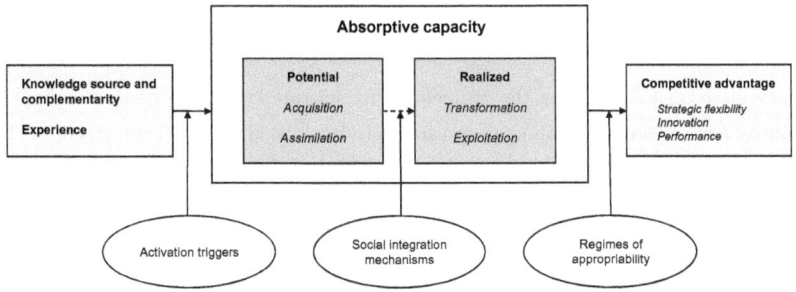

**Figure 2.6** A Model of Absorptive Capacity (Source: Zahra & George 2002, p. 192)

The advantage of this perception of absorptive capacity is that the construct is not under-stood as a firm's static and stable character but as a process that consists of important steps that complement each other. Additionally, applying the RBV explains differences in firms' performances as a function of different levels of vigorousness in each of the four dimensions. Although absorptive capacity is a strong concept to understand what capabilities are required to use external knowledge efficiently, regarding the full spectrum of open innovation, its scope is insufficient as it focuses on an outside-in perspective only. Thus, additional capabilities are sought after in order to fully manage innovation outside the boundaries of the firm.

Lichtenthaler & Lichtenthaler (2009) developed a framework that extends the concept of absorptive capacity by combining prior research into open innovation, dynamic capabilities, and ambidexterity, that is, research about the internal balance between exploration and exploitation (O'Reilly III. & Tushman 2011, Raisch, Birkinshaw, Probst & Tushman 2009). By focusing on an integrative perspective, they responded to the deficiencies of absorptive capacity as an outside-in flow since this refers to an external exploration, internal retention, and internal exploitation only. However, other paths of knowledge flows are possible. For example, internally developed and maintained knowledge may be exploited externally (Lichtenthaler 2007). In particular, the framework thus consists of two dimensions, namely knowledge flow and locus of knowledge. By adopting the process-based view of absorptive capacity, Lichtenthaler & Lichtenthaler (2009) identified that exploration, retention, and exploitation of knowledge may be organized internally and externally. Each combination of point in knowledge flow (exploration, retention, exploitation) and locus of knowledge (internal, external) calls for specific knowledge capacities. Consequently, absorptive capacity, which refers to exploring external knowledge, is complemented by five other capacities, namely inventive, transformative, connective, innovative, and desorptive capacity (Figure 2.7).

Inventive capacity denotes the competence to generate knowledge inside the firm. For example, any endeavor pertaining to research activities would fall into this category. Thus, it may be considered the internal counterpart of absorptive capacity. Transformative capacity is needed to manage the retention of knowledge within the firm. This includes maintaining the firm's knowledge base, by, for instance, knowledge sharing in order to stay capable of acting when people leave the firm and take their knowledge with them. Transformative capacity, therefore, refers to what prior research understood as (internal) knowledge management (Hansen et al. 2005). Connective capacity is defined as a firm's

| | Knowledge exploration | Knowledge retention | Knowledge exploitation |
|---|---|---|---|
| **Internal (Intrafirm)** | Inventive capacity | Transformative capacity | Innovative capacity |
| **External (Interfirm)** | Absorptive capacity | Connective capacity | Desorptive capacity |

**Figure 2.7** Capability-Based Knowledge Management Capacity Framework (Source: Lichtenthaler & Lichtenthaler 2009, p. 1318)

ability to retain knowledge in interfirm relationships. This definition is somewhat confusing as alliances often aim to co-develop new knowledge. However, considering the framework, creation of external knowledge would refer to absorptive capacity. Lichtenthaler & Lichtenthaler (2009, p. 1320) dissolved this theoretical conflict as follows: "In contrast to absorptive capacity, external knowledge retention does not assume inward knowledge transfer. Instead, firms ensure privileged access to external knowledge without acquiring it." In other words, connective capacity is required to abide existing relationships to external partners.

The latter two capacities refer to the exploitation phase I already discussed in the context of knowledge spillovers in section 2.2.2. Whereas innovative capacity refers to a firm's competence to exploit knowledge internally, desorptive capacity denotes a firm's ability to externally commercialize knowledge (Lichtenthaler & Lichtenthaler 2010). To avoid confusion, it is important to clarify that in both scenarios the knowledge will be commercialized. The difference, however, is that firms that have built or acquired the knowledge commercialize the knowledge itself, whereas external exploitation is realized by means of out-licensing. Regarding the process-based component of the framework, both capacities are disconnected from the origin of the knowledge. In other words, for the exploitation phase, it is irrelevant if the knowledge was created or maintained internally or externally. Admittedly, prior related knowledge is an inalienable component in each of the six knowledge management capacities.

Considering the compatibility of absorptive capacity as defined by Zahra & George (2002) and absorptive capacity as defined by Lichtenthaler & Lichtenthaler (2009), several interferences and contradictions must be highlighted. First, absorptive capacity according to Lichtenthaler & Lichtenthaler does not include the realization part of Zahra & George's definition. In particular, Zahra & George (2002, p. 190) refered to realized absorptive capacity as "a function of the transformation and exploitation capabilities [...] Realized absorptive capacity reflects the firm's capacity to leverage the knowledge that has been absorbed," a function that is reserved for innovative and desorptive capacity in the Lichtenthaler & Lichtenthaler framework.

Second, regarding the retention phase, transformative capacity has to be distinguished from transformation according to Zahra & George. At first glance, both concepts refer to similar aspects of knowledge management; however, there is a clear difference. While Zahra & George's transformation capability reflected the process of combining existing and newly acquired and assimilated knowledge, Lichtenthaler & Lichtentahler's transformative capacity refered to the perpetuation of knowledge exclusively.

Third, although the Lichtenthaler & Lichtenthaler framework covered the full range of knowledge activities, namely knowledge exploration, retention, and exploitation, Zahra & George provided a more detailed description (e.g., components of knowledge acquisition and assimilation). As a result, in the work of Zahra & George, the process character became more perceptible in favor of a rather limited scope, that is, focusing on outside-in knowledge flows only.

To sum up, by complementing absorptive capacity Lichtenthaler & Lichtenthaler (2009, p. 1322) developed the concept of knowledge management capacity that is defined "as a firm's ability to dynamically manage its knowledge base over time by reconfiguring and realigning the processes of knowledge exploration, retention, and exploitation inside and outside the organization." In contrast to prior research, the framework provides an integrative view on capacities needed to manage permeating boundaries of the firm and thereby extends the open innovation paradigm as well as the absorptive capacity construct.

## 2.3.3 Extending Knowledge Management Capacity for Open Innovation

Absorptive capacity, alone or within the larger framework of knowledge management capacity, has been identified as a critical component for firms to arrive at a competitive position. Thus, the concept clearly stands in the tradition of the RBV and KBV that posit that a firm derives competitive advantage by means of superior resources (cf. Barney 1991, Grant 1996b, Tsai 2001). Various researchers have acknowledged that firms can derive competitive advantages not just from the resources they embody themselves but also from resources external to the firm but to which the firm can secure access (cf. Mathews 2003). However, as noted by Walsh, Schaarschmidt & Von Kortzfleisch (2011), most of these studies focus on external resources a firm has to pay for to obtain.

In a similar vein, Zahra & George (2002) as well as Lichtenthaler & Lichtenthaler (2009) did not explicitly include free external sources in their arguments. However, knowledge acquisition may have pecuniary and non-pecuniary components (Dahlander & Gann 2010) and to acquire external knowledge does not necessarily mean to pay for that knowledge. For example, as discussed in Section 2.2.1, users, which may be individuals or firms alike, may provide their knowledge for free. In addition, firms that use an OSS development approach extent their knowledge base by internalizing external knowledge they not have paid for. Thus, including a pecuniary component into the concept of knowledge management capacity provided by Lichtenthaler & Lichtenthaler (2009) seems to be reasonable.

Furthermore, as noted by Lichtenthaler (2011c), knowledge management capacity refers to an organizational-level and is a function of project-level decisions and individual-level attitudes. This view is consistent with prior work on organizational learning where a firm's knowledge and learning capacity is commonly thought of as a result of individuals' and project teams' or units' learning performance (e.g., Amin & Cohendet 2000, Gong, Huang & Farh 2009, Remedios & Boreham 2004, Tsai 2001). Concerning decisions on the project-level, firms have the option to chose between *make* and *buy* in the exploration phase (Cassiman & Veugelers 2006), between *integrate* and *relate* in the retention phase, and between *keep* and *sell* in the exploration phase (Lichtenthaler & Ernst 2006). However, at the individual-level, various personal attitudes influence firm decisions at the project level and therefore the embodiment of knowledge management capacity at the firm-level. Consequently, each form of knowledge management capacity, namely incentive, absorptive, transformative, connective, innovative, and desorptive capacity is affected by specific individual attitudes.

These attitudes may have positive and negative consequences. Regarding the NIH-syndrome, which denotes employee behavior against external knowledge exploration (e.g., Katz & Allen 1982), although an overemphasis of internal knowledge creation inhibits additional benefits, it also may lead to an invigorated internal R&D department. For example,

SAP was long well known for its rejection of external developments; however, their rather closed approach was very successful. The counterpart of the NIH-attitude is described as "buy-in" attitude, that is, individuals have an attitude toward externally exploring knowledge (Lichtenthaler 2011c). Again, this attitude may have positive and negative effects. Whereas additional gains may be realized by acquiring external knowledge, deploying resources to acquire external knowledge may limit the internal R&D performance. An overemphasis on buying external knowledge may therefore cause negative effects. Generally, however, research on R&D related attitudes at the individual-level predominantly focus on the negative sides. Lichtenthaler (2011c) completed the individual-level view on his framework by discussing employee attitudes in the context of knowledge retention and exploitation, namely "not-connected-here", "relate-out", "not-sold-here", and "sell-out", and further admitted that some attitudes are relatively unlikely at present, such as the "sell-out" attitude, which would point to an overemphasis of externally exploiting knowledge (Lichtenthaler, Ernst & Hoegl 2010).

Admittedly, a single individual's attitude does not affect every decision on the project-level nor does a single decision on the project-level determine a firm's knowledge management capacity. In addition, no firm would be claimed to have an underdeveloped absorptive capacity because it has decided to build the required knowledge for a single new technology on its own. However, expended prevalence of a negative attitude may (over time) affect the firm's knowledge management capacity. In turn, as attitudes are often a result of the organizational culture (e.g., Martins & Terblanche 2003), it equally is possible to limit a negative attitude's diffusion (or its results, respectively) within the firm by introducing an adequate corporate culture.

Consequently, as reciprocal interactions between individual-, project-, and firm-level are affected by employees' attitudes, I argue that the framework has to be extended by introducing a pecuniary dimension. In particular, I argue that at the individual-, project-, and firm-level, attitudes, decisions, and capabilities differ according to the pecuniary or non-pecuniary nature of knowledge. For example, Schaarschmidt & Kilian (2011) showed that employees at a telecommunication company are more skeptical towards free external knowledge than towards external knowledge the company has paid for. In a similar vein, decisions at the project-level are not limited to make-or-buy decisions (cf. Cassiman & Veugelers 2006), as the firm may chose between pecuniary (e.g., buying patents) and non-pecuniary (e.g., reading customers' blogs) solutions for acquiring external knowledge. Regarding knowledge exploitation, the firm's options are not limited to sell out unused knowledge. For example, Netscape made the source code for the Netscape Navigator, an Internet browser, publicly available due to market share losses to its competitor, Microsoft's Internet Explorer. The source code then was taken by a community of individual developers and became the browser known as Mozilla Firefox.

Thus, if a firm's knowledge management capacity is determined by successive decisions at the project-level and decisions, in turn, by attitudes at the individual level, I argue that capabilities required to absorb, maintain, and exploit external knowledge a firm has to pay for differ from capabilities needed to deal with knowledge which is available for free. In order to capture the pecuniary dimension of knowledge management capacity, I merge components of Dahlander & Gann's (2010), Lichtenthaler & Lichtenthaler's (2009), and Lichtenthaler's (2011c) work to provide an extended framework that is used for this thesis (Figure 2.8).

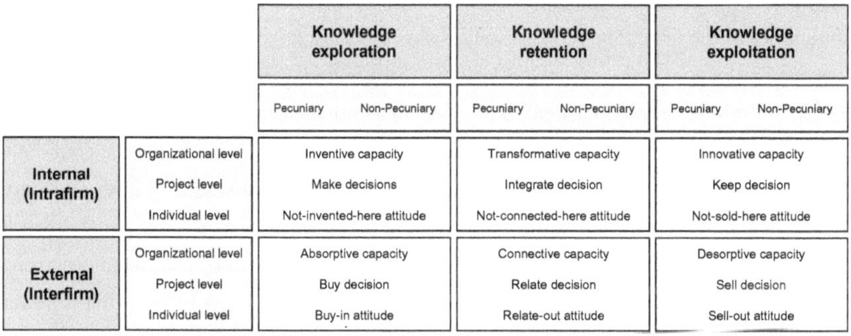

| | | Knowledge exploration | | Knowledge retention | | Knowledge exploitation | |
|---|---|---|---|---|---|---|---|
| | | Pecuniary | Non-Pecuniary | Pecuniary | Non-Pecuniary | Pecuniary | Non-Pecuniary |
| **Internal (Intrafirm)** | Organizational level | Inventive capacity | | Transformative capacity | | Innovative capacity | |
| | Project level | Make decisions | | Integrate decision | | Keep decision | |
| | Individual level | Not-invented-here attitude | | Not-connected-here attitude | | Not-sold-here attitude | |
| **External (Interfirm)** | Organizational level | Absorptive capacity | | Connective capacity | | Desorptive capacity | |
| | Project level | Buy decision | | Relate decision | | Sell decision | |
| | Individual level | Buy-in attitude | | Relate-out attitude | | Sell-out attitude | |

**Figure 2.8** Extended Capability-Based Knowledge Management Capacity Framework

In particular, each phase of the Lichtenthaler & Lichtenthaler (2009) framework is extended by a pecuniary dimension at the individual-level, project-level, and firm-level. In line with Dahlander & Gann (2010), pecuniary refers to transactions in which money is involved in exchange whereas non-pecuniary refers to indirect benefits. Admittedly, the differences between these dimensions are predominant regarding external knowledge exploration, retention, and exploitation and not very apparent regarding internal knowledge. However, as especially large firms increasingly install internal clearing systems, internal knowledge transactions are no longer only non-pecuniary. Moreover, it is possible to charge another organizational unit for a service, or knowledge transfer, respectively.

In summary, by including a pecuniary and non-pecuniary dimension, the knowledge management capacity framework has been extended. This is important in order to provide a theoretical basis for this dissertation and to capture aspects of the OSS phenomenon, which will be discussed in the following chapters. In addition, as the extended framework sketches the entire continuum of capabilities needed to organize for open innovation, it may act – like the original framework – as a basis for future empirical research in open innovation.

## 2.4  Summary

The aim of this chapter was (1) to define the R&D boundaries of the firm, (2) to describe the increasing dissolution of firm boundaries and the open innovation phenomenon, and (3) to provide an extended capability-based framework for managing innovation beyond the boundaries of the firm. The RBV as well as new institutional economics deliver illustrative reasons for the emergence and the dilution of firm boundaries that help in understanding the open innovation phenomenon. The open innovation phenomenon, in turn, may best be explained as a function of increased search for external knowledge on the one hand and increased knowledge spillovers on the other. The challenges that derive from open firm boundaries and knowledge acquisition and spillover in particular, require capabilities at the firm-level, which, in turn, are influenced by decisions at the project-level and personal attitudes at the individual-level. Lichtenthaler & Lichtenthaler (2009) provided a powerful framework that may be applied in multiple scenarios. However, with regard to this dissertation and its research objective a pecuniary dimension had to be included in order to differentiate between free external knowledge and knowledge a firm has to pay for to obtain.

Although the extended framework was developed explicitly for this dissertation, it may serve research streams such as dynamic and combinative capabilities (e.g., Kogut & Zander 1992), ambidexterity (e.g., Raisch et al. 2009), or knowledge management in general (e.g., Hansen 2002). In this regard, it may especially help to classify newly emerged phenomena, such as user-generated content or active customer integration, within the more sustained research on technological product development.

# Chapter 3

# Commercializing and Controlling Open Source Software Development

## 3.1 Exploring the Open Source Phenomenon

This chapter captures a boundary spanning position between Chapter 2, which was intended to outline the challenges firms are confronted with by opening their boundaries, and Chapters 4 and 5. To fulfill its designated function, the phenomenon of firm participation in the development of OSS will first be presented against the background of managing innovation beyond firm boundaries. Second, to deepen the understanding of firm participation, the role of business models and their underlying basic economic principles will be highlighted. Third, building upon a distinctive understanding of value creation and value capture, specific characteristics of OSS will be expanded to a systematization of OSS business models. However, the commercialization of existing OSSDPs is likely to disrupt natural governance structures of OSS communities (O'Mahony & Ferraro 2007). In particular, as firms bring their interests to the table, formerly convergent interests among participants within the OSS community are supplemented by firm interests. Consequently, interests within a community increasingly become divergent. In the worst case, divergent interests between voluntary community members and firms let voluntary developers deallocate their resources from the project, which, in turn, destroys the benefit of outside innovation. Finally, therefore, by relying on the understanding of OSS communities as boundary organizations (cf. O'Mahony & Bechky 2008), I discuss different options firms may apply to influence and control a project's trajectory and thereby extend established control theories pertaining to classical organizational forms.

## 3.1.1   A Brief Introduction to Open Source Software Development

Although there are many definitions of what constitutes OSS, the basic idea behind OSS is very simple: A software qualifies as OSS if it is released under a license proved by the Open Source Initiative (OSI) (Deek & McHugh 2007). In order to align with OSI requirements, a software license must possess the following characteristics:[1]

1. Free redistribution

2. (Availability of the) source code

3. (Possibility to make) derived works

4. Integrity of the author's source code

5. No discrimination against persons or groups

6. No discrimination against field of endeavor

7. Distribution of license

8. License must not be specific to a product

9. License must not restrict other software

In practice, free redistribution of the human-readable source code (instead of a compiled version) and, even more importantly, (free) redistribution of modified versions and derived work determine differences between open source and proprietary software. Here, "free" first and foremost refers to free access to the derived work, not necessarily to "free of charge". Due to the openness and accessiblity of the source code, everyone is enabled to install OSS, may look at how functions are programmed, and modify the installed versions. Users thereby become producers in that they actively make changes to their version of the source code, which is allowed as long as they do not exclude others from using and scrutinizing their derived work.

Problems, changes, and requests for future functionalities are then discussed online since OSSDPs use web and other Internet sites as the repository for the source code, documentation, and discussion (Lakhani & Von Hippel 2003). In its purest form, all

---

[1]The Open Source Definition (Annotated), Version 1.9, http://www.opensource.org/osd.html, last access 6/8/2011

contributors to an OSSDPs are volunteers and thereby form what is commonly known as "the community." Thus, belonging to a community or not is not formalized: "Basically, whoever shows up is the team" (Goldman & Gabriel 2005, p. 32). However, users may equally be firms or individuals employed by firms. If these persons follow their firm's interests, usually a more formalized structure emerges within a community (Dahlander & Wallin 2006). Thus, since for every OSSDP a seemingly unique community exists, not all OSSDPs are the same.

In addition, other terms, such as "free software" (FS) and "free and open source software" (F/OSS) are commonly used interchangeably with OSS. Thus, in order to draw a coherent picture of firm engagement in OSS development, OSS has to be distinguished from other forms of free software. "Free software" is the oldest form of non-proprietary software. The term was created in 1984 by Richard Stallman as a response to an increasing number of proprietary software development at that time. He believed that software should be free and open generally and therefore founded the Free Software Foundation (FSF) in 1985 to support his beliefs. The FSF fosters four types of "freedom" (cf. Alexy 2009), namely:

- The freedom to run the program, for any purpose,

- the freedom to study how the program works, and adapt it to your needs,

- the freedom to redistribute copies so you can help your neighbor, and

- the freedom to improve the program, and release your improvements to the public, so that the whole community benefits. Access to the source code is a precondition for this.

These freedoms were implemented in licenses such as the GNU General Public License (GPL). Yet, only a few software packages run on a stand-alone basis. The majority of software has to be either connected to other software or, especially in commercial use, combined with and integrated in other proprietary software. When using GPL, if OSS under a GPL license is combined with third-party proprietary software, the proprietary software has to be released under the GPL as well. Since this is a seemingly unattractive attribute for commercial use – firms would have to open their protected proprietary software – with GPL, firms are kept away from OSS development. In turn, as noted by several authors (e.g., O'Mahony & Bechky 2008, Stewart et al. 2006), firm participation may be crucial for a project's survival as firms provide OSSDPs with necessary resources and infrastructure. Whereas the freedoms as defined by Stallman and the FSF restrict commerical use and particpation, the definition provided by the OSI allows for licenses that are more business friendly, such as the Berkely Software Distribution (BSD) that will be discussed in the successive section. Thus, the main difference between OSS and FS lies in the non-applicability of reciprocal usage or as Alexy (2009, p. 11) highlights: "Whereas software under the GPL or similar licenses would both be considered OSS and FS, several OSS licenses do not guarantee as much freedom as to call the respective software free software."

## 3.1.2 Granting Access to the Use of Intellectual Property by Licensing

Generally, software consists of a collection of IP, which, in turn, is regarded a result of mental activity that has to be protected against unauthorized use (Gambardella & Hall 2006). As previously discussed, copyrights and patents are forms of IP protection that exclude others from the usage of IP created by the copyright owner. However, in a commercial sense, a copyright owner has no interest in keeping others away from using his IP completely; he only wants to ensure that no party benefits from using his IP without benefiting himself. In other words, a copyright owner is likely to grant access to his IP if he receives (monetary) compensation for this access. In turn, this form of exchange is regulated by contracts.

Licenses are specific kinds of contracts as "a license is a contract between the owner of a product and a prospective user that grants the user certain rights regarding the use of the product that would otherwise be illegal" (Deek & McHugh 2007, p. 232). In particular, the copyright owner for a source grants permission to individuals or groups to use what the copyright protects. A license then defines the terms and conditions under which the licensee (i.e., the person or group who wants to use the IP of others) is allowed to use the IP of the copyright owner. For example, in the case of software, a user usually has to agree to certain conditions, such as not to distribute the software or not to produce illegal copies, prior to the actual usage of the software.

Considering software as a collection of IP, it is important to note that granting access to the usage of IP by means of licensing does not imply transfering ownership of that IP. The copyright owner can still limit the granted access to specific functions of the software or even withdraw access if license agreements have been violated. This literally means that a buyer of, for instance, a Microsoft Word distribution does not own the software itself – although the impression might arise as buyers are provided with physical components such as a packaging or a DVD. Confusingly, although a customer only pays for the license (i.e., he pays for the usage only), typically he has to agree to a license's conditions *after* he has bought the physical components in a store.[2]

OSS licenses differ from other forms of software licenses in terms of restriction of usage and the possibility to screen the source code or not (Koski 2005). For example, in the case of freeware and shareware, the program is only delivered in its compiled form. Therefore, users are excluded from looking into the source code. In addition, in the case of shareware, the free usage is usually limited to a certain timeframe. Once this time has expired, the user has to buy a proprietary version of that software if he intends to continue his usage. Figure 3.1 provides an overview of different software licenses and their characteristics.

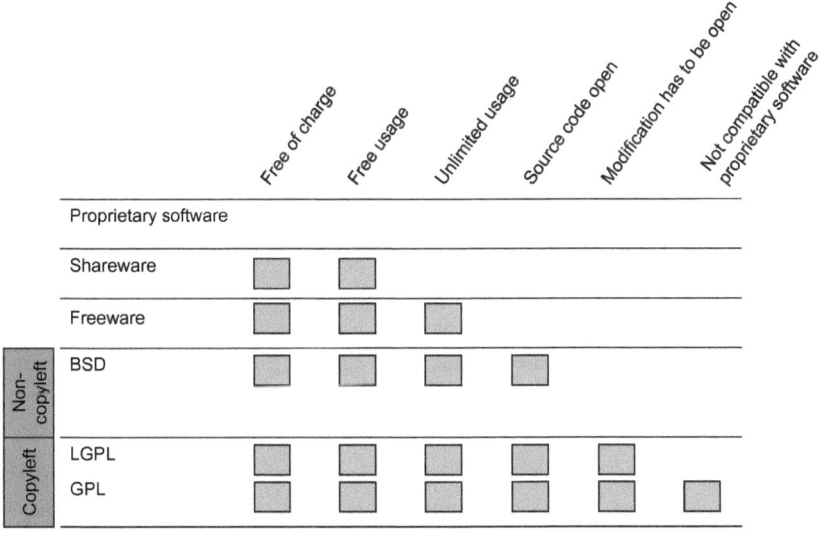

**Figure 3.1** Overview of Types of Licenses and their Characteristics

**Copyleft Licenses.** With the development of OSS, two types of licenses are distinguished, namely copyleft and non-copyleft licenses. As the choice of a license determines the applicability of business models, knowing about the range and limits of these licenses is crucial for successfully managing OSSDPs.

---

[2]I thank Dr. Martin Braun, lawyer at WilmerHale, Frankfurt, for clarifying this point during a workshop organized by the Open Source Business Foundation (OSBF) in Frankfurt, February 2009.

The term *copyleft* is a wordplay derived from copyright, which implies the "right to copy", and moreover to modify and distribute IP. In contrast, copyleft licenses do not secure rights to IP by the principle of exclusion but by ensuring that each distribution of an author's work has to stay freely accessible. Consequently, an author will always have the possibility to access his work. Although this statement sounds quite tautological at first glance, considering the possibility of copyright transfer, the copyleft principle might better secure an author's right than the copyright principle because a future copyright owner could exclude the creator from using the IP. Admittedly, the economic value of IP is not protected as with copyleft licensing, third parties are either able to access the work (McGowan 2005). Such software that is not longer protected by copyright in its purest sense is said to be in the public domain (Deek & McHugh 2007). In addition, software that was once licensed under the terms of a copyleft license is excluded from being published under a conventional license in the future.[3] Since copyleft licenses cannot ignore legal copyright law, it requires two steps to provide software under the terms of a copyleft license:

> To copyleft a program, we first state that it is copyrighted; then we add distribution terms, which are a legal instrument that gives everyone the rights to use, modify, and redistribute the program's code or any program derived from it but only if the distribution terms are unchanged.[4]

The most prominent copyleft license is the GNU General Public License (GPL) provided by the FSF. GPL was modified to version 2.0 in 1991 (GPLv2) and to version 3 in 2007 (GPLv3) as a response to changes in digital rights management (DRM) systems such as technical copy protection. Linux, probably the most famous OSSDP, runs on a GPL basis.

Elementary for copyleft licenses are the four freedoms as defined by Stallman and the FSF that are based on the (ideological) tenet to provide better software to the people by joint development. In order to prevent free riders from selling extensions to the source code or the source code itself, software that contains GPL licensed parts must be provided under the GPL again (e.g., De Laat 2005, Fershtman & Gandal 2007, Kennedy 2004, St. Laurent 2004). This restriction corrodes firms' ability to take the source code, modify it, and sell the enlarged version as proprietary software.[5] However, this also means that

---

[3]In particular, the owner of the entire copyright to software always has the possibility to provide different licenses (i.e. including rights to use the software). Therefore, the copyright owner might provide a software with different, non-copyleft licenses in the future. However, he cannot withdraw the rights he already has granted with the copyleft license.

[4]"What is copyleft?", URL, http://www.gnu.org/copyleft/copyleft.html, last access 6/10/2011

[5]In popular perception, OSS is always free of charge. Although this is true *de facto*, *de jure* one can sell

if firms include GPL code in their (probably much bigger) code for proprietary software, they would have to distribute the entire software as GPL software. Consequently, GPL and other copyleft software licenses are not very business friendly – and were intentionally never designed to be.

Less restrictive licenses, such as the LGPL or the Eclipse Public License (EPL), explicitly allow for software code to be integrated in OSS under different licenses and in proprietary software. For example, the LGPL, an acronym for Less General Public License, pays attention to the fact that software libraries, which are designed to be used in different programs, may in particular be implemented in proprietary software without infecting the proprietary code with GPL.

**Non-copyleft Licenses.** In contrast to pure copyleft licenses, non-copyleft licenses allow for manipulation and redistribution of OSS without the obligation to open the derived work or to dictate the type of license. Consequently, components of software under a non-copyleft license may be integrated in proprietary software without making the derived product publicly available. Although these types of licenses do not share the ideological tenet of early OSS, they are also approved by the OSI, such as the Berkeley Software Distribution License (BSD). Thus, non-copyleft licenses still require offering the source code, but support commercial interests (McGowan 2005).

### 3.1.3 Open Source Software and Intellectual Property Management

Copyright as understood in Anglo-Saxon countries differs from the German "Urheberrecht." Whereas the former originates from a utilitarian approach to IP, the latter is based on laws of nature (Dreier & Nolte 2006). One tenet of the laws of nature is that IP is automatically accepted *as is* in the moment it was created. From that moment on, the "Urherberrecht" protects the IP of its creator up to 70 years after his death (Zirn 2004). In addition, according to a monistic perspective, the law aims to protect both, ideological and economic interests. In contrast, until 1976, in the US the creator of IP had to apply for registration of the copyright (Carroll 2007). Furthermore, the US copyright aims more to protect economic investments rather than the ideological value of IP. The copyright

---

OSS code. This is not directly restricted by the OSS licenses. However, as OSS is a public good such as water or air, the likelihood that someone will pay for something that is available for free anyway is limited. Nevertheless, by adding value to the code (e.g., new functionalities), software vendors might mobilize new customers.

system as known today has its roots in the British publishing sector. The thought behind protecting the economic value of IP, though, originates from the copyright act, known as "Statute of Anne" from 1710, which conceded copyrights formerly accorded to publishers to authors (Khong 2006).[6] With the accedence of the US to the Berner convention in 1989, the US copyright law was adapted. Today, the copyright protects a person's IP from the moment it was created until 70 years after the person's death, the same as the German Urheberrecht (Zirn 2004).

However, one important basic difference between Anglo-Saxon copyright law and the German Urheberrecht remains. In the German system, copyrights to IP are not transferable – except by inheritance (Koglin & Metzger 2004). Thus, the applicability of IP transfer is restricted to the transfer of rights to use the IP. Ownership, though, remains with the creator of IP. Within the Anglo-Saxon law system, it is possible to transfer the entire IP rights, including ownership, which embraces the right to further transfer the IP and the right to grant access to third parties to use the IP (Carroll 2006).

This produces a number of important implications for the development of OSS. OSS usually is developed by a distributed group of developers, some paid by a firm while others voluntarily contribute to the OSS project. Therefore, speaking of *the* copyright raises the wrong assumption that copyright to an OSS is a monolithic thing. Moreover, OSS consists of many different parts and modules, developed by different programmers, and thus protected by different copyrights. Since OSSDPs cannot ignore legal copyright law, the management of an OSSDP has to ensure that the project possesses copyrights to *every* module before it can give the code to the public.

In principle, three approaches to deal with copyright are possible (Fogel, Barkhau, Menge & Pittinger 2010). First, project management could ignore legal issues. This is not a recommendable strategy since legitimate copyright holders may deny offering their parts of the software to be publicly available at any time. Second, project management might obtain licenses to entire parts of the software. Developers are then asked to sign a *contributor license agreement* that basically is a declaration of agreement that gives the project the right to use the respective parts and put it in the public domain. Third, developers can transfer the copyright to their current and future contributions to the project. This is the most legally admissible solution as the project becomes the owner of the entire copyright to the software.

---

[6]As the British Statute of Anne did not apply to the American colonies, it took until 1790 before US law adopted the principle of shifting rights to IP from the publisher to the author. Therefore, in some sources, the first *copyright act* refers to the year of 1790.

However, with different law systems, the third way is not practicable either if the project is hosted or the developer has produced the parts of the software in a non-copyright law country (e.g., Germany). In addition, contributions by firms are considered critically. Firms that intend to provide software to an OSSDP equally have to ensure that they possess the copyright to that software. However, employees usually yield their rights to their IP to the firm for any invention they make during their working time or by using firm resources. Yet the majority of employment contracts do not consider the possibility to open source the software. In particular, the employee has only agreed to use his IP for the generation of products whose copyrights ought to remain with the firm. They have never explicitly agreed to give their IP to the public domain.[7]

Additionally, software provided by firms usually contains proprietary third-party components that have to be removed before the firm can either contribute the code to an existing project or initiate an OSSDP itself. For example, many proprietary components had to be removed from many modules of what is today known as the Swordfish project, hosted on the Eclipse website. Considering these legal issues, building a business model entirely dependent upon OSS is very shaky.

### 3.1.4 Open Source as Resource Allocation Beyond Firm Boundaries

Early research on OSS has focused on why people voluntarily contribute to OSSDPs, a question that not only attracted information systems researchers but psychologists and economists alike (e.g., Harhoff, Henkel & Von Hippel 2003). Numerous taxonomies of motives and interests exist by which incentives may be categorized (Hertel 2007). For example, by drawing on the concept of signaling to future employees, Lerner & Tirole (2002) distinguished immediate from delayed incentives. However, the majority of research on developers' motivations has drawn upon the distinction between intrinsic and extrinsic motivation (e.g., Hars and Ou 2002, Hertel, Niedner & Herrmann 2003, Lakhani & Wolf 2005, Roberts et al. 2006, Shah 2006). According to the seminal work of Ryan & Deci (1980, 1987), intrinsic motivation refers to being driven by the wish to satisfy basic human needs. Tasks then become interesting and will be pursued without the prospect of additional rewards (Deci 1971). In contrast, extrinsic motivation is considered to be a response to an external obligation (Deci, Koestner & Ryan 1999).

---

[7]I again have to thank Dr. Martin Brown for highlighting this important point.

Following Deci & Ryan (2000), the interplay between intrinsic and extrinsic motivation is not as static as it might seem at first glance. By introducing the concept of self-determination (Deci & Ryan 1985), the authors explain that each dimension (i.e., intrinsic and extrinsic motivation) may be intentionally internalized or blocked. For instance, although a mother forces a child to clean up her room (extrinsic motivation), the child might respond with a non-execution of the task. In a similar vein, extrinsic incentives may be turned into intrinsic ones. Even though externally motivated, the person might pursue an activity in the same manner as if he has been intrinsically motivated. For example, need for social embeddedness, an extrinsic incentive, may be internalized. Deci & Ryan (1987) used the term *internalized extrinsic motivation* to refer to situations similar to the latter example.

Considering the development of OSS, various reasons have been identified for voluntary contributions. The most prominent examples embrace fun in problem solving, enjoyment of coding (e.g., Shah 2006), identification with the community, improvement of software code (e.g., Lakhani & Wolf 2005), and reputation in the community (e.g., Muller 2006). In their attempt to expose relationships between different types of motivation, Roberts et al. (2006, p. 985) stated: "Contributing to OSS projects for the sheer enjoyment of coding is clearly an intrinsic motivation whereas being paid to contribute is quintessential extrinsic motivation."

They further highlighted that other types of motivation ask for more careful classifications. Enhancing status or career opportunities by contributing to an OSSDP is extrinsic by definition but may be better categorized as internalized extrinsic motivation. In one result of their study among Apache developers, they suggested that communities should "largely welcome commercial efforts by companies" (Roberts et al 2006, p. 997). This is a very surprising finding as previous work delineated the picture of the community as a closed circle of voluntary contributors who aimed to form an antipole to proprietary software development and who explicitly excluded firms from participation. In particular, Roberts et al. (2006) showed that extrinsically motivated programmers have lower use-value motivation (i.e., motivation to satisfy their own needs) while they possess greater status motivation (i.e., motivation aiming to increase reputation within the community). Additionally, being paid is considered as a means to produce a higher level of contributions to the source code.

Thus, not only firms may benefit by working with a community but the community might benefit from firm participation as well. However, this reciprocal relationship is very sensitive since OSS communities are based on trust, fairness, and reciprocal gift exchange (Bergquist & Ljungberg 2001, Lakhani & Von Hippel 2003, Shah 2006). Consequently, community members expect firms to equally provide resources, knowledge, or contributions to the code base as a form of social exchange (Brown & Duguid 2001, Méndez-Durón & Garcia 2009). The assumption that with an OSS development approach a firm gets access to thousands of developers for *free* therefore remains an anecdotal myth (Goldman & Gabriel 2005). Admittedly, what is true is that this engagement does not necessarily lead to direct costs. Instead, given the reciprocal nature of gift exchange, firms willing to join an existing OSSDP or even to initiate a project have to provide advance performance. For example, Walsh et al. (2011) showed, based on social exchange theory (SET) (Homans 1958) that firms can influence the provision of free resources external to the firm by assigning a higher number of paid programmers to the project.[8] In a similar vein, Grand, von Krogh, Leonard, & Swap (2004, p. 592) argued that "engagement with OSS community is a highly dynamic and cumulative process of gift exchange.".

The mutual OSS development of firms and voluntary coders has challenged established models of technological innovation. In particular, OSS development is neither a complete *private innovation*, that is, returns to the firm result from private goods, nor *collective invention*, that is, firms collaborate with public institutions to produce a public good (Grand et al. 2004). Instead, both volunteers and firms leverage their private resources to produce a public good that is characterized by its non-excludability, its non-rivalry, and the fact that the value of the good does not decrease through multiple consumption (Alexy 2009). Von Hippel & Von Krogh (2003) referred to this anomaly in innovation theory as *private-collective innovation*. Figure 3.2 sketches the pathways private investments may take in various innovation modes.

However, as Bitzer, Schrettl & Schröder (2007, p. 161) argued, "yet, in general, economic theory would predict that privately provided public goods suffer from problems of under-provision, delays in supply, and inferior quality." Unlike this prediction, OSS features remarkable success and in some areas even outperforms proprietary products (Rossi Lamastra 2009), a fact that increasingly attracts firms to invest in OSS development.

Given that communities tend to guard their commons (O'Mahony 2003), applying an OSS approach to software development, or more generally, to open innovation is by far more complex than it might seem. Indeed, firms are interested in getting access to free

---

[8]Social exchange involves a series of interactions that generates obligations between entities. Therefore, SET explains exchange as a function of reciprocal obligations instead as a function of economic transactions (Cropanzano & Mitchell 2005).

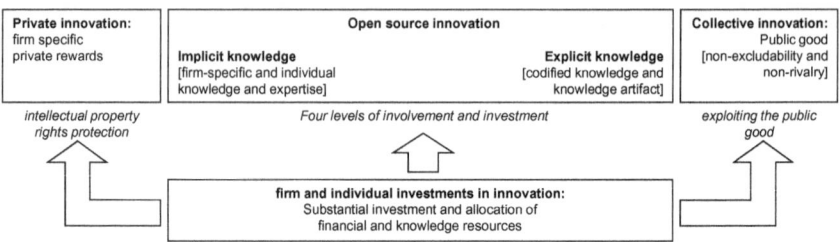

**Figure 3.2** Resource Allocation to Open Source Innovation (Source: Grand et al. 2004, p. 594)

external labor and thereby perform outsourcing without directly paying for that service (Ågerfalk & Fitzgerald 2008). Moreover, although community members are often referred to as hobbyists, they build a pool of highly skilled developers. For example, Hars & Ou (2002) reported that more than 70% of the community members has a bachelor's, master's, or PhD degree. Being able to establish access to this resource pool may therefore leverage a firm's competitive advantage at no direct costs.

Grand et al. (2004) proposed a four-level model for resource allocation beyond firm boundaries (Figure 3.3). In order to get familiar with OSS and to exploit opportunities, in level one, the firm is a user of OSS. Using OSS is not completely costless since the firm either has to build the technological capabilities to integrate OSS into the corporate IT infrastructure or pay for this service. In fact, as no license fees accrue, total cost of ownership is considered to be lower compared to proprietary software.

Hardware producers especially may use OSS as complementary assets in level two. Here, the OSS becomes an integral part of a physical product. For example, many handheld phones run on the Linux operating system (Henkel 2007) or Linux derivatives, such as Android. In contrast to level one firms, level two firms do not only use the software as an end-user, but integrate OSS in their own products and distribute them. This option is not limited to hardware producers, but usage with other types of software sometimes is restricted by OSS licenses, such as GPL. As the effort to integrate OSS in their own products exceeds the effort to align OSS with an internal IT environment, firms must invest more in internal development and knowledge capabilities than level one firms.

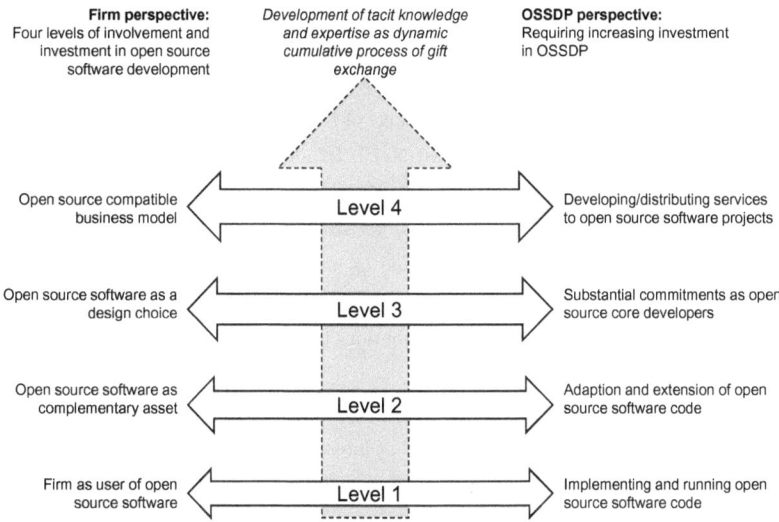

**Figure 3.3** A Four-level Management Model of Resource Allocation (Source: Grand et al. 2004, p. 595)

In level three, firms adopt the OSS development approach for selected projects as a design choice. For example, Innotek, a German software company, chose to offer its virtualization software Virtualbox as OSS and changed the development process accordingly before it was acquired by Sun Microsystems. Employing OSS as a design choice requires the adoption of certain OSS principles and process models, such as releasing early and often and switching to more agile programming methods (e.g., Extreme Programming) (Gassmann, Sandmeier & Wecht 2006). In addition, as software is barely developed from scratch, adopting the OSS approach to software development is very cost intensive. For example, firms that intend to use the OSS development style and open their software to the public have to put effort into modularizing the software code and removing third-party components.

At level four, OSS "moves from being a design choice for particular projects to the design choice for a firm's overall business model" (Grand et al. 2004, p. 599). If a firm does not start at level four – which principally would be possible – this movement implies a major business model innovation (Chesbrough 2010). For example, since OSS is usually free of charge, firms at level four are likely to migrate to service companies and deliver support and maintain a customer's software solution. Although these companies still produce software code, considering themselves a service company is demanding for employees.

Additionally, if employees were used to their closed development approach, changing to an OSS business model requires the change of their mindsets to avoid NIH or not-sold-here problems (cf. Lichtenthaler & Ernst 2006). In addition, establishing an OSS friendly corporate culture is very cost intensive. RedHat is considered a groundbreaking example for a level four company since its revenues first stemmed from selling CD ROMs to assist the Linux installation and now are realized by maintaining OSS server infrastructures. However, at level four, a company has an increased responsibility to the community that expects substantial and high quality contributions to the OSSDP. Figure 3.3 illustrates the differences among the four levels.

In summary, as participants' performance is a function of motivation, knowledge, skills, and abilities (Roberts et al. 2006), firms that intend to work with a community must first know motivation and incentive structures of participating developers. Moreover since OSS communities work on concepts such as gift exchange, participating firms must provide resources to the project. Firms therefore may get access to complementary free external resources without directly paying for to obtain them, but they have to provide resources themselves, especially if they migrate to higher levels of OSS approaches.

# 3.2    Exposing Characteristics of Business Models

Although a consensus of the relevance of business models for turning latent value into countable rents is prevalent in both academia and practice, the concept of business models is surprisingly a rather diffuse one (Magretta 2002). The term first came up in 1957 in an article written by Bellman, Clark, Malcolm, Craft & Ricciardi. Since then, numerous researchers have provided possible definitions of what might constitute a business model, especially since the year 2000 (cf. Osterwalder, Pigneur & Tucci 2005), but those definitions covered such a broad range that they explain everything and nothing. Regarding the aim of this chapter, to analyze the options by which firms may commercialize OSS, it is important to know what a business model is and, even more importantly, what it is not. Therefore, the following section aims at providing an understanding of a business model that will build the basis for commercialization approaches pertaining to OSS.

### 3.2.1 Similarities and Disparities of Business Model Definitions

Despite the undisputed importance of business models for the purposive execution of business tasks, surprisingly, a resilient definition is missing. Various definitions of business models have been provided in recent years but only a few are commonly accepted. The prevalence of the term can be traced back to the new economy boom at the beginning of the century. Even though the idea of business models is much older, Internet start-ups that relied on financial investors used the term especially to articulate their uniqueness as they lack classical arguments for fundraising such as unique patents and existing infrastructure (Schubert & Hampe 2006).

In accordance with Timmers (1998, p. 4) who delivered one of the most cited definitions, a business model is "an architecture for the product, service and information flows, including a description of the various business actors and their roles; and a description of the potential benefits for the various business actors; and a description of the sources of revenues." He further added that a business model is usually complemented by a marketing model, which is equally important.

Chesbrough (2010) went a little further and provided a definition that combines and extends what Timmers (1998) identified as a business and marketing model. Building upon his own work with Richard Rosenbloom (2002), Chesbrough (2010, p. 355) considered business models to fulfill a set of functions because they:

- Articulate the value proposition (i.e., the value created for users by an offering based on technology);

- Identify a market segment and specify the revenue generation mechanism (i.e., users to whom technology is useful and for what purpose);

- Define the structure of the value chain required to create and distribute the offering and complementary assets needed to support a position in the value chain;

- Detail the revenue mechanism(s) by which the firm will be paid for the offering;

- Estimate the cost structure and profit potential (given value proposition and value chain structure);

- Describe the position of the firm within the value network linking suppliers and customers (incl. identifying potential complementors and competitors); and

- Formulate the competitive strategy by which the innovating firm will gain and hold advantage over rivals.

Based on the work of Brandenburger & Stuart (1996), Gambardella & McGahan (2010, p. 263) described a business model as an "organization's approach to generating revenue at a reasonable cost, and incorporates assumptions about how it will both create and capture value." In the definition provided by Teece (2010b, p. 173), it became obvious that a business model is not necessarily a result of intentional planning – a fact which may be used to separate business models from strategies:

> Whenever a business enterprise is established, it either explicitly or implicitly employs a particular business model that describes the design or architecture of the value creation, delivery, and capture mechanisms it employs. The essence of a business model is in defining the manner by which the enterprise delivers value to customers, entices customers to pay for value, and converts those payments to profit. It thus reflects management's hypothesis about what customers want, how they want it, and how the enterprise can organize to best meet those needs, get paid for doing so, and make a profit.

In their book for a management audience, Osterwalder & Pigneur (2010) focused on nine core elements that are indispensable for business models:

1. Customer segments

2. Value proposition

3. Channels

4. Customer relationship

5. Revenue stream

6. Key resources

7. Key activities

8. Key partnerships

9. Cost structure

Likewise, in another paper, Osterwalder et al. (2005, p. 17) focused on the importance to understand the concept of business models as a function of a set of elements and their relationships. Accordingly, they provided a more goal-oriented definition and emphasized the relevance of business models for revenue generation:

> A business model is a conceptual tool that contains a set of elements and their relationships and allows expressing the business logic of a specific firm. It is a description of the value a company offers to one or several segments of customers and of the architecture of the firm and its network of partners for creating, marketing, and delivering this value and relationship capital, to generate profitable and sustainable revenue streams.

In addition, in German-speaking countries, the business model concept provided by Wirtz (2008) has been widely adopted. He differentiated six different connected models that together constitute the business model, namely, marketing model, procurement model, value creation model, value offer model, distribution model, and revenue model.

Based on the few definitions of business models presented above, it may be recorded as an intermediate result that both value creation and value capture are vitally important to define the firm's business model. Moreover, it is conspicuous that either explicitly or implicitly, all definitions call for a revenue stream or a revenue model (e.g., Amit & Zott 2001, Chesbrough & Rosenbloom 2002, Teece 2010b). At first glance, this finding might not be surprising as firms inherently aim to profit monetarily from their engagements. However, regarding various commercialization approaches to OSS, which will be discussed in section 3.3, this finding highlights that many forms of firm engagement in OSS development may not be regarded as a business model. For this dissertation, in accordance with Alexy (2009), a business model is considered the way by which the firm creates and delivers value for its customers and how the firm appropriates rents from that value.

Additionally, to avoid confusion in the use of various terms, business model has to be delineated from strategy. Although numerous situations exist, in which a one-to-one mapping between strategy and business model is apparent, more recently, a business model is argued to be a reflection of the firm's realized strategy (Casadesus-Masanell & Ricart 2010). I disagree as this would imply that a firm would change its business model anytime it changes or adds a strategy. For example, cost leadership (Porter 1996), which may seemingly be considered a strategy, does not necessarily affect the firm's business model, that is, the way the firm generates value and provides it to its customers. I further found support for my suggestion in the work of Gavett & Rivkin (2007, p. 423): "For us, a strategy is a management team's way of seeing its place in its environment as well as the firm's way of interacting with the environment." As they do not use the terms revenue model or value creation, concepts that have been identified as vitally relevant for the firm's business model, I suggest an alternative explanation for the relationship between strategy and business model. For this research, strategy is distinguished from business model by its level of abstraction. Whereas firms are generally well advised not to employ multiple business models, a firm must have multiple strategies to address different challenges.

### 3.2.2 Modularity in Technologies, Organizations, and Business Models

Pisano & Teece (2007) highlighted the shift of many industries from vertical to horizontal industry architectures. They referred to the computer industry to demonstrate how modular technological architectures have not only affected the architecture of physical products but the architecture of industries alike. These changing conditions impact mechanisms of value creation and value capture. For example, in less horizontal industries, firms design, manufacture, and distribute their inventions themselves. In contrast, with increasing verticalization, or industrialization, respectively, those activities are distributed among different firms. As this provides many avenues for innovative business models, firms must understand the concept of modularity, both on a firm and an industry level. Thus, the theoretical roots of the concept of modularity will be illustrated next.

Organization theorists have a long tradition of investigating the relation between (complex) product design and the organization that develops it (Henderson & Clark 1990, Marple 1961, Ulrich & Eppinger 2004). While some researchers found product architecture and organizational structure to be misaligned (Sosa, Eppinger & Rowles 2004), others found that a product's design tend to mirror the structure of the organization that designed the product (Conway 1968, MacCormack, Rusnak & Baldwin 2006, Sanchez & Mahoney 1996). However, although there is contradictory evidence on organization and product structure interactions, it is widely accepted that modularity, both in terms of organization and product design, is an important factor when dealing with (complex) architectures.

Architectures of complex products are typically decomposed into systems and components (Baldwin & Clark 2000, Brusoni & Prencipe 2011, Henderson & Clark 1990, Van Schewick 2005). If the interdependence within one component is high and the interdependence between components is (relatively) low, then the structure (of a product or an organization, respectively) is considered to be modular (Baldwin & Clark 2000, Langlois 2002). Following the predominant logic provided by various researchers, modular design structures are particularly useful when systems become large and difficult to handle (Ethiraj & Levinthal 2004, MacCormack, Baldwin & Rusnak 2010, Parnas 1972). To further investigate this assumption, recent research has focused on OSSDPs as a field of study as both communities of developers as well as architectures of the source code are considered to be complex (Langlois & Garzarelli 2008, MacCormack et al. 2006, MacCormack, Rusnak & Baldwin 2008). OSS is characterized by a number of aspects. However, as stated in section 3.1, at its heart it is (1) the distribution of a program's human-readable source code and (2) a licensing model that allows everyone to modify and further develop the product, which separates OSS from proprietary software (Von Hippel & Von Krogh 2003). These aspects usually lead to distributed (virtual) communities of developers collectively working on a project. In order to reduce the level of complexity inherent in an OSSDP, software architects favor a modular design, which becomes especially visible if OSS is compared to a proprietary counterpart developed within the boundaries of a firm (MacCormack et al. 2006).

However, not only is OSS code modular due to the complexity of the existing organization that develops the code – a group of individual programmers usually distributed over the globe – but due to the fact that modularity is a prerequisite to attract new developers to work for a project (Milev, Muegge & Weiss 2009, Raymond 1998, West & Lakhani 2008). As the entire source code is not easy to understand as a whole, prospective participants in an OSSDP usually start their engagement by looking at small decomposed (sub-)systems of the product. From a theoretical point of view, one can observe patterns of Gidden's (1979, 1984) theory of structuration and the duality of structure in two dimensions. First, the group of distributed developers is modular per se - and needs to be modular to overcome the task's complexity. This circularly-recursive duality is consistent with Gidden's (1984, p. 64) assumption that "human actions are enabled and constraint by structures." Second, the software code is modular due to the structure of the group that developed it and needs to be modular as a prerequisite to secure the engagement of new-to-the-project programmers.

Drawing on the literature mentioned above, I consider different business models to lead – or that at least ought to lead – to different degrees of modularity in the software's architecture. For instance, as shown by MacCormack et al. (2006), code that was produced within the boundaries of a firm is more likely to have an integrated design. According to Sosa et al. (2004, p. 1675), "in the development of highly complex products, it would be naïve to expect a perfect mapping between design interfaces and team interaction." In a similar vein, I suggest that business model and code structure are not perfectly aligned. However, following this logic, business models designed to attract external participants to work for a project are more likely to show a higher degree of modularity than those developed in-house by a software vendor.

Thus, regarding this research, modularity may be a vitally important concept to explain different business models in general and to explain business models related to the commercialization of OSS in particular.

### 3.2.3   Platforms and the 'Razor-Razor Blade Model'

As noted by Teece (2010b, p. 174), "the concept of a business model has no established theoretical grounding in economics or in business." This stems partly from a theoretical culture within economics by which customer behavior has been predominantly argued as being dependent on price and market structure. Intangible products and services have thereby barely been considered. However, although business models have no place in economic theory (Teece 2010b), some revenue models, which are considered an integral part of the firm's business model, are based on economic mechanisms.

For example, the "razor-razor blade" model is a prominent business revenue model based on pricing razors inexpensively while expensively pricing the complementary good (i.e., the razor blade) (Mahadevan 2000). This model is very generic and has been widely adopted, such as in the case of printers and cartridges. Furthermore, this model not only applies to physical products but may be extended to be applicable for service-product bundles. As services deliver a more permanent revenue stream, often the product is offered at a low price, while the service is priced extensively. However, as services are not protectable by IPRs, competition in the service part is likely to increase. New market entrants might build on products that they not have developed themselves and offer their services at lower costs or better quality. For example, after the purchase of the car, the customer does not necessarily have to call on the dealer's garage service but is free to get his car maintained at an independent car repair shop (Chesbrough 2011).

This extended form of the basic model may also be applicable in situations where pricing the good is difficult. For example, information is known for being very difficult to price as its value does not decrease with multiple consumption, but may decrease over time.

Hence, newspapers are sold at a price that approximately covers the costs of the production process, but for covering the remaining costs and for obtaining profits, publishers rely on advertising revenue. Consequently, the price of a newspaper does not mirror the value of the information it contains (Chesbrough 2010). In a similar vein, TV stations deliver the broadcast stream to end consumers for free while demanding remuneration from firms that want to advertise their products. Admittedly, the prices for advertising may vary in accordance with the content. During Formula 1 races or the Super Bowl, TV stations tend to demand dramatically higher prices. Firms seeking to place advertising are likely to pay higher prices as they get access to millions of potential customers in return. This definitely represents a value to the advertising firm. In turn, the value of TV stations is a function of its coverage, which itself is a function of the number of reachable individuals and quality of content. In particular, the more viewers are mobilized, the more attractive the TV station becomes for advertising companies (Teece 2010b).

Firms that employ such business models are considered to run a platform business model. Platform business models are an increasingly adopted way to transform technological innovations into economic value (Teece 2010b). A platform is defined as an infrastructure that allows two distinct markets to interact with each other (Boudreau 2008). For example, Sony Playstation is a platform as it combines the market for end users and the market for game developers. End users buy the platform (e.g., Sony Playstation) and buy applicable games. They would never buy the platform without games or games without the platform. In order to speed up the platforms diffusion, Sony Playstation is sold below its producing costs. Game developers, however, have to pay for licenses to develop their games for Sony's Playstation. Thus, without the platform, end users and game developers cannot interact.

Recently emerged forms of platform business models do not even demand a monetary compensation for the platform itself, as the benefit from network effects preponderates (Boudreau 2010). Facebook, probably the most famous example for a platform business model, allows free developers as well as software firms to develop their own applications for the platform. Instead of keeping the platform closed, Facebook accepts leaving possible revenues to external parties in favor of a rapid diffusion of the platform as well as increased traffic. Eisenmann, Parker & Van Alstyne (2008) therefore discussed when and how platforms should be opened in order to attract more users *and* more usage. The value of a platform such as Facebook lies in the number of people who spend their time on the platform (Parker & Van Alstyne 2008). Recently, Groupon launched a platform by which consumers are enabled to buy merchant coupons. As the merchants do not pay for being presented on that platform, there is no cash flow from the vendor to the platform. However, the platform pockets a portion of the cash that is intended for the coupon provider.

These examples portray business models that are based on the mechanisms of two-sided markets and two-sided network effects (Parker & Van Alstyne 2005). Their advantage is predicted to rest upon the fact that network effects leverage rapid diffusion. These principles are equally applicable for the commercialization of OSS.

### 3.2.4    Hybrid Value Creation and Capture With Business Models

Recent research in economics has emphasized the distinction between value creation and value capture (Lepak et al. 2007, Pisano & Teece 2007). In times of open firm boundaries it also has been highlighted that value creation and value capture (i.e., value appropriation) may be distributed among different players in a market. Consequently, the one who has created the value is not necessarily the one who obtains profits in the end (Teece 2003).

However, the logic has predominantly applied to products. Hybrid value creation, that is, the value is created by products *and* services, has gained momentum in recent years. Whereas in the industrialization phase, revenue streams almost entirely came from the sale of products, today, a large proportion of the gross domestic product (GDP) in modern economies comes from services or product-service bundles (Chesbrough 2010). This is challenging firms that had a business model with a focus on the sale of products. Chesbrough (2011) described the shift producers of cell phones had to conduct in order to respond to the demand of services allocated with mobile devices. For example, Motorola had been very successful with its product Razr V3 but failed to provide adequate services. In contrast, competitors provided services such as game and song downloads by which the cell phone could be personalized through allocated stores (e.g., Android Markets, Apple App Store, Nokia OVI Store) (Hampe 2010). As a consequence, in the case of cell phones, the buy-decision is no longer exclusively dependent on the quality and looks of the product. Chesbrough (2011) argued that firms that miss out on transforming their business to services have failed to innovate their business model towards services.

Within the marketing literature, this trend toward services is referred to as *service-dominant (S-D) logic* (Vargo & Lusch 2004). In particular, the term suggests that even if people buy goods, what they actually buy is a service that fulfills their needs. As the great Harvard marketing professor Theodore Levitt used to tell his students: *"People don't want to buy a quarter-inch drill. They want a quarter-inch hole"* (cf. Christensen, Cook & Hall

2005). In consequence, hiring a craftsman who is being paid for the service of drilling the hole would lead to the same result – but would make buying the drill obsolete. However, although services have evolved to a preponderance in modern economies in recent decades, the original intention behind S-D logic was not to emphasize that an increase in services is observable (Vargo & Akaka 2009, Vargo & Lusch 2008, Vargo, Maglio & Akaka 2008). Instead, Vargo & Lusch (2008, p. 4) intended to provide a concept that explains economic exchange on services rather than goods:

> S-D logic says that the application of competences for the benefit of another party – that is, service – is the foundation of all economic exchange. Thus, even when goods are involved, what is driving economic activity is service – applied knowledge.

By applying the perspective of hybrid value creation and value appropriation, I emphasize the following four revenue models by which firms may turn economic value in revenue streams. Again, firms may apply different revenue models to different products without affecting their business model. The following approaches therefore indeed mirror revenue models rather than business models.

Regarding product-service bundles, either the product or the service can be offered for free (entirely or for a certain time) while the complementary asset will be priced. In particular, for the purpose of rapid diffusion, it is very common to grant access to specific offerings for free, while demanding remuneration after people got familiar with the offering. Many business social software portals have used this approach. Access to the platform was granted for free, but to be able to use all functionalities, the customer had to become a member with monthly payments. This model is referred to as "freemium", composed out of "free" and "premium" (Teece 2010b). In a similar vein, Skype, an Internet-based telephone service is offered for free. However, the customer is charged for reaching participants in the old public switched telephone network (PSTN) and for using extended features.

The concept of giving something away for free in order to benefit in the long run has also been applied to OSS. As depicted, OSS is a public good and free of charge. The integration of OSS into existing IT environments, however, requires appropriate knowledge and capabilities. Many firms therefore have been founded with a business model dedicated to offer services for installing, integrating, and maintaining existing OSS. For instance, since various content management systems (CMS) are available as open source, such as Drupal, Joomla, or Modx, firms providing complementary services have evolved that position themselves as solution providers.

By offering some drinks for free at the beginning, bar keepers may generate customer loyalty with the result that people stay longer or come back more often. In a similar vein, wine tasting often is charged at cost price. In particular, the wine for the wine tasting is priced inexpensively – in the case of exhibition even offered for free – along with the hope to sell an increased number of wine bottles afterwards. In Psychology, this underlying mechanism is referred to as foot-in-the-door technique (Freedman & Fraser 1966). Table 3.1 provides an overview of different mechanisms for revenue streams based on free provision at the outset.

**Table 3.1** Systematization of Revenue Streams

| Free in $t_1$ | $ in $t_2$ | Characteristic |
|---|---|---|
| Service | Product | Rare, as revenue stream is less interrupted in cases of services |
| Service | Service | Freemium model, basic service free (e.g., Skype) |
| Product | Service | Rapid diffusion (e.g., OSS) |
| Product | Product | Extreme case of razor-razor blade model (e.g., wine tasting) |

As these examples show, giving something away for free in a first round in favor of selling more in a second round is an elaborated, but also very old-established mechanism (e.g., wine tasting). Firms that employ such mechanisms benefit from the awakening of unperceived demands and rapid diffusion in the case of products. Additionally, business models that are dependent on network effects especially make use of free revealing, such as in the case of Skype that has to provide the first customers with a number of reachable contacts – also known as "critical mass" – to be able to offer their service. Thus, firms that reveal services or products to their customers for free have nothing to donate. Instead, they use the mechanisms described above to acquire new customers and to increase their speed of growth.

## 3.3 Commercializing Open Source Software Through Business Models

By discussing different options of value creation and value appropriation as well as free revealing for rapid diffusion and hybrid value creation (i.e., value coming from products and services), it has been shown that firms can benefit from giving something away for free. The same mechanisms apply for OSS. Thus, it is possible to benefit from open source software by using, contributing, and releasing (Alexy 2009). The following subsections therefore aim at providing a systematization of commercialization approaches and business models around OSS.

### 3.3.1 Commercial Versus Community Open Source Software

The commercial production of OSS has attracted a lot of attention in recent years (Dahlander & Magnusson 2005, Fosfuri et al. 2008). Success stories like Linux, MySQL, or JBoss have proven that nowadays OSS has the quality and the customer acceptance to compete with its proprietary rivals. However, although the term open source suggests that software that claims to be OSS shares a coherent body of attributes, at its core, the only connecting attributes are (1) delivery of the source code in a human readable form and (2) a license approved by the Open Source Initiative (OSI). Based on these two factors, many development styles or commercialization approaches are possible, which, although very different in terms of motivation and goals, are considered to be open source (Raymond 1999). For example, even Microsoft, a candidate for high quality proprietary software products, has released open source licenses, such as Microsoft Reciprocal License (Ms-RL), which are consistent with OSI requirements.

Recent research further confirms that commercial OSS exists in many different ways according to its revenue model, type of license, development style, number of participating firms, number of participating volunteers, or governance mode (Bonaccorsi et al. 2006, Dahlander & Magnusson 2008, West 2003). Consequently, core business functions like community management, sales, marketing, product management, engineering, and support differ among different commercialization strategies (Watson, Boudreau, York, Greiner & Wynn 2008).

In order to categorize an increasing variety of commercialization approaches, authors basically distinguish commercial from community open source (e.g., Capra & Wassermann 2008, Riehle 2009, Riehle 2011a). In accordance with these publications, control and ownership structures are the critical indices in order to differentiate between these two types. While community OSS is controlled by a community of stakeholders (including multiple individual programmers and/or firms), commercial OSS is controlled by exactly one stakeholder with the purpose of commercially exploiting it. In addition, with commercial OSS, the software is literally owned by the software vendor while IP to community OSS is widely distributed among stakeholders (see Section 3.1.2 and Section 3.1.2). As I will discuss later, owning the entire software code is crucial for a number of commercialization approaches. Some authors have referred to projects with one sole dominating company as single-vendor projects (SVP) (Riehle 2011b). Consequently, projects that consist of more than one active company are called multi-vendor projects (MVP)(Schaarschmidt, Bertram & Von Kortzfleisch 2011). Whereas single vendor approaches show similarities to proprietary software vendors' behavior, in cases of multiple firms active in a project, development is being processed as in R&D alliances or joint ventures (Schaarschmidt & Von Kortzfleisch 2009). In the latter case, usually a direct revenue stream is not intended. Instead, multiple firms often combine their resources in order to build a platform and to promote standards with the aim to sell on top applications along with complementary products or services.[9]

Dahlander (2007) pointed to the fact that a project's history is equally important to categorize commercialization approaches. Building upon a sound case study research, he provided a systematization combining degree of firm participation and project initiation. In particular, he focused on *de novo* entrants and argued that OSSDPs either may be initiated by a firm or a community. A *de novo* firm is defined as "an independent for-profit organization, whose core business is to utilize expertise in FOSS to develop products or services. [...] The rationale for focusing on de novo entrants is that these firms were founded specifically to commercialize FOSS" (Dahlander 2007, p. 919). By applying this definition, established firms active in OSS such as IBM, Sun, or Oracle have been excluded from this analysis. The resulting framework is provided in Table 3.2.

**Table 3.2** Typology of Commercialization Approaches (Source: Dahlander 2007, p. 930)

|  | Firm initiated | Community initiated |
|---|---|---|
| High degree of firm participation | Approach I | Approach II |
| Low degree of firm participation | Approach III | Approach IV |

Each entry approach is composed of a set of different characteristics. Approach I describes scenarios where firms have founded an OSSDP by revealing source code to the public and continue to contribute to that project. Although similarities to SVPs are observable, Dahlander (2007) did not exclude the possibility that multiple firms reveal code simultaneously to an OSSDP. The challenge of that approach is determined by the absence of a

---

[9]It is worth noting that the definition of community OSS is not very precise. It assumes that the community may consist entirely of voluntary contributors. Although such projects exist, they are not of interest here. In the case of OSS, the term *community* therefore includes for-profit actors and voluntary contributors.

community a priori. Firms that intend to use Approach I have to put effort on the creation of an active community consisting of voluntary contributors and other firms. In Approach II, firms harness existing communities. Since OSS communities work on a gift-exchange basis (Bergquist & Ljungberg 2001), this approach requires giving something back to the community to prevent a bad reputation. Thus, firms sponsor communities by providing infrastructure or skilled employees that are assigned to the project.

Regarding Approach III, that is, firm initiation and low degree of firm participation, firms may give source code to the public, but do not actively support future development. This is often the case in situations where firms are unable to capture value from their developments because of inferior market position. For example, Netscape has offered its Browser Navigator as OSS – a product that became the Mozilla Browser – as a response to the market dominance of Microsoft's Internet Explorer. After revealing the code, an active developer community evolved around the product without any effort by Netscape.[10] Finally, Approach IV embraces situations in which peers founded their own OSSDPs based on existing ones devoid of firm participation. These projects are either not attractive for firms or actively prohibit firm engagement. Table 3.3 summarizes the four approaches Dahlander (2007) has identified.

By developing this classification of commercialization approaches, Dahlander (2007) implicitly developed a classification of possible OSS business models – although he used the term very rarely. Furthermore, he followed the assumptions that business models have to be related to revenue streams and thus explicitly excluded Approach IV from being a business model (Table 3.3). Considering Approaches I-III, although rationales for firm participation and engagement are provided, their interference prevents the provision of a more detailed picture. In other words, the framework facilitates endeavors to classify OSSDPs and to explain basic mechanisms but fails to provide business models a firm could directly apply. In the following, I will thus put attention to important aspects such as licensing and product structure with regard to business models.

---

[10]To be precise, Netscape did not support the project after it went public. However, as previously discussed, before proprietary software can be turned into OSS, firms often have to reprogram the code in advance. This is necessary to provide source code to the public that is modular enough to attract voluntary developers and free of third-party components.

Table 3.3 Overview of Potential Commercialization Approaches (Source: Dahlander 2007, p. 931)

| | Approach I Firm initiated – high degree of firm participation | Approach II Community initiated – high degree of firm participation | Approach III Firm initiated – low degree of firm participation | Approach IV Community initiated – low degree of firm participation |
|---|---|---|---|---|
| License practice | Firm choose licensing scheme | Community initiator chooses licensing scheme | Firms chose licensing scheme | Community initiator chooses licensing scheme |
| Community development | Firms initiate the community and try to build a critical mass through devoting resources to build the community | Peers initiate the community and over time it becomes valuable for the firm because it develops something new or at low cost | Firms initiate the community and try to build acritical mass but fails or only release source code that is not considered critical for the firms business | Peers initiate a community but firms do not participate because it is not of relevance or they are fended off by the members |
| Business model | Attempt to build a business model closely knit to the developments of the community and thus have a strong incentive to influence community development | Attempt to build a business model closely knit to the developments of the community and thus have a strong incentive to influence community development | Firms release source code and hope to get adoption without actively building the community | Not applicable because firms are not actively participating and trying to build a business model on the community |
| Protection of returns | Influence the direction of the community by devoting personnel to work as peers Keep some control of the interactions in the community | Influence the direction of the community by devoting personnel to work as peers Actively screening communities for new developments and integrate if beneficial | | |

## 3.3.2 Benefits from Using, Contributing to and Revealing Open Source Software

As Alexy (2009) has pointed out, firms can benefit from OSS by using, contributing to, and revealing OSS. Using OSS embraces the unmodified installation, implementation, and integration of OSS into the firm's IT infrastructure. Alexy (2009, p. 25) explicitly included situations where firms use OSS to complement their own developments without contributing to the OSSDP. Contributing to existing OSS refers to situations in which the company provides its own extensions and modifications of an OSS to the original OSSDP. The people in charge of the project then decide if such a contribution will be included in a future release or not. Finally, releasing proprietary software as OSS is an option by which firms give their own privately developed software code to the public for free.

Regarding the use of OSS, the benefits are quite obvious at first glance. In contrast to highly priced licenses for proprietary software, the firm gets the software for free. Furthermore, as it is possible to screen the software code a priori, firms can check if the new software fits their existing IT infrastructure or not. However, this is only half the truth. What really counts is the total cost of ownership (TCO), that is, the sum of hardware costs, software costs, and costs for support and maintenance. Hardware costs, support, and maintenance costs are not affected by the use of OSS. Using OSS therefore reduces TCO, but does not eliminate them. According to Augustin (2008), using OSS may reduce TCO about 30% on average in medium sized and large IT environments.[11] The success of IT-dependent firms such as 1&1, Amazon, or Google can be (partly) traced back to the usage of an OSS infrastructure through which they could grow faster.

As with any new software, OSS has to be integrated in the firm's existing IT landscape. This embraces tasks such as connecting the new OSS to mail servers, building connections to internal databases, and including OSS in the firm's security strategy. In order to keep the sophisticated IT environment running, new OSS has to be customized to specific customers' needs (Keßler & Alpar 2009). Yet, customization often involves changes in the source code. Firms usually avoid customizing the source code itself as they fear non-compatibility with the next release of the software, but sometimes customizing the source code is inevitable.

---

[11] Alexy (2009) additionally pointed to several downsides of using OSS. In particular he addressed search costs as usually an OSS will not meet the firms need perfectly. Therefore, firms have to search themselves for the ideal solution to their problems or even have to combine different OSS products. This is not congruent with my observations. IT firms hardly search for solutions themselves. Usually consulting companies offer the best solution for the company's need, mostly consisting of proprietary and OSS components. As this is no longer OSS specific, I do not include this as a serious OSS downside here.

Firms that have made use of OSS intensively have therefore implicitly gained OSS specific skills and new solutions. With OSS, there is only a thin line between being a user and being a developer.

Contributing to OSS is possible in many ways (Jullien & Zimmermann 2009). In contrast to popular perception, not everyone is allowed to contribute to the project's source code. Access to the source code of an elaborated OSS is restricted to a small group of people who have shown their development skills in advance. For example, the Linux kernel was developed by Linus Torvalds and his "trusted lieutenants," a handful of selected developers (Shaikh & Cornford 2003). When people speak of thousands of developers who contribute to an OSS, they usually refer to minor tasks such as reporting bugs or requesting new functionalities (Goldman & Gabriel 2005). Consequently, firms might benefit from contributing to OSS in several ways.

Firstly, if firms intensively report bugs, it is most likely that the OSS community will provide a new update to fix bugs – a fact that distinguishes OSS from many proprietary software vendors. Secondly, the firm can benefit from providing an internally developed extension to the community. In particular, if the firm has developed an extension to the OSS that serves the needs of various users, it is likely that this extension will be included in a future release. In turn, the firm receives a future version that is perfectly compatible with its infrastructure. Again, this procedure is often referred to as gift exchange (Bergquist & Ljungberg 2001). Admittedly, competing firms will benefit as well from the extension. Finally, the firm may benefit from revealing, that is, giving away in-house developed (proprietary) software. As discussed previously, this option is mostly used for strategic reasons, such as in the case of Netscape Navigator and Mozilla (see Section 3.3.1).

Many researchers have emphasized that it is hardly – if at all – possible to (economically) benefit from OSS directly (e.g., Alexy 2009, Dahlander 2005, Fitzgerald 2006). However, in contrast to popular perception, it is possible to sell OSS. This is not restricted by most of the licenses. Yet, as with other forms of public goods such as water or air, people do not tend to pay for offers they can get elsewhere for free. For example, beverage companies respond to the non-excludability of water by offering added value, such as admixing minerals or serving supermarkets with bottled water. In addition, if the firm is able to communicate the water's purity through effective marketing – easy in times where trust in official water pipes is decreasing – people are likely to spend a multiple of the water's actual price. Similar principles are applicable to OSS.

Yet, direct revenue streams are very seldom feasible with OSS. Thus, OSS is mostly used to facilitate alternative and complementary revenue streams. Building on the benefits from using, contributing to, and releasing OSS together with basic principles of hybrid value creation and capture (see Section 3.2.4), I will provide an overview of the most common OSS business models and possible side effects in the following section.

### 3.3.3 Open Source Software and Business Models

The previous sections have shown that, generally, as long as there is no commonly accepted systematization of business models, there will not be a reliable systematization of OSS business models. A recent study by the 451 Group even highlighted that OSS is a development model rather than a business model.[12] However, there are economic principles underlying the development of OSS by which a business model can be realized. The following thoughts therefore first sketch economic mechanisms pertaining to OSS. Afterwards, I conclude by providing a basic classification of OSS business models.

#### 3.3.3.1 Basic Economic Mechanisms Pertaining to Open Source Software

By comparing OSS to open innovation, West & Gallagher (2006) distinguished four major types of OSS commercialization based on knowledge flows, namely pooled R&D, spinouts, selling complements, and donated complements.

Pooled R&D is close to what other bodies of literature call R&D alliances or consortia (O'Mahony & Vecchi 2009, Vanhaverbeke et al. 2002). In the latter case, at least two firms have put together their R&D effort to develop products according to their needs (Arranz & de Arroyabe 2004). Recent research shows that firms not only cooperate to save costs but also in cases of strong vertical relationships and high technological uncertainty (West & Gallagher 2006). Moreover, firms donate money and development capacities to open source projects in order to boost the sales of related products. Typically, this kind of OSS donation concentrates on software designed for lower levels of the OSI reference model, like operating systems or web servers. The generation of revenue is realized through sale of applications built on top. Furthermore, using OSS in the sense of pooled R&D helps to define open standards and to avoid incompatibility in inter-firm collaboration. However, pooled OSS R&D differs fundamentally from typical R&D consortia in the non-controllability of evolved

---

[12] " Open Source Is Not a Business Model", http://www.the451group.com/caos/caos_detail.php?icid=694, last access: 6/22/2011

knowledge and in the challenge of integrating contributions from external participants that are partly unpaid.

In some cases, technology is no longer useful for a firm, especially when it does not contribute to the firm's performance or does not fit into the corporate strategy. Firms then stop maintaining a product and leave it to those who developed it. Freed from firms' stranglehold, these spinoffs oftentimes are able to perform well as observable in the case of Xerox and Adobe (Chesbrough 2003). If a firm continues to support the technology, it is possible to remain control and generate demand for related products or services. West & Gallagher (2006) called this kind of OSS approach "spinout" (Figure 3.4).

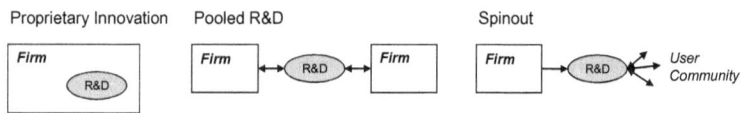

**Figure 3.4** Knowledge Flows in Three Software R&D Models (Source: West & Gallagher 2006, p. 323)

Selling complements means that firms are giving away innovations for free to generate revenue from complementary goods (Boudreau 2008). This reflects the razor-razor blade model discussed previously. Razor blades as well as ink cartridges generate more revenue than the printer or the razor, respectively. In the hardware-software paradigm (cf. Katz & Shapiro 1986), firms use two-sided pricing strategies and lose money on the hardware to profit from the complementary good (i.e., software). In the case of separated target groups such as the video game or web server industry, firms address consumers as well as developers. A two-sided pricing strategy can increase the overall profit (Gallaugher & Wang 2002). In turn, firms can offer the software at a price near zero while profiting from hardware. For example, IBM has donated software code as well as personnel to the Apache Web server foundation, and, more importantly, has abandoned the development of its own Web server (West & Gallagher 2006). Although no exact figures exist, IBM has benefited from the rapid diffusion and usage of the OSS Web server Apache and was able to sell more server hardware to serve this diffusion.

Closely related to the concept of selling complements is the concept of donated complements. Although firms do not receive direct revenues neither through the core product nor by selling complementary goods, they can focus on customer loyalty by opening their products for user adjustments. As it is not possible to gain from the openness directly, firms create an infrastructure to invite people to participate and hoping for increased publicity and reduced development costs (West & Gallagher 2006).

West & Gallagher (2006) provided a summary of the rationales that drive firm engagement in OSS. Yet, they were also very cautious to use the term *business model*. Considering the business model definition for this research, that is, the way by which the firm generates and delivers value for its customers and how it appropriates the rents from that value, it is not surprising that West & Gallagher (2006) were cautious in using the term business model since some approaches simply do not meet the definition. For instance, using OSS as a means to build R&D consortia is not a business model as long as it is not clear where the rents come from. As firms seemingly may have different interests to participate, pooled R&D itself is not a business model. However, firms could integrate the pooled R&D approach in a superordinated business model.

### 3.3.3.2 A Classification of Open Source Business Models

As the preceding explanations show, providing a classification of OSS business models is challenging for numerous reasons. Firstly, if no direct revenue stream will be obtained from OSS, any possible OSS business model automatically builds on the sale of complementary products and services. This implies that all approaches that intend to classify OSS business models should be subcategories of selling complementary goods and services. Secondly, the boundaries between using, contributing to, and revealing OSS are dissolving. Thirdly, the terms using, contributing, and revealing have to be understood as a continuum from using, providing, and revealing source code occasionally to intensely and exclusively using, contributing, and revealing OSS. For instance, with regard to contributing to OSS, different contributing behaviors mirror different strategies and thus may not be subsumed under one single OSS business model. Fourthly, today many proprietary software products by software vendors consist of proprietary and open source parts that make the delineation of OSS from proprietary even more difficult. Finally, one must not forget that not only software vendors are active in OSS development. Hardware producers and firms active in IT intensive industries, such as financial services, equally engage in OSS development but their motives to participate are not comparable.

Considering the various obstacles to classify OSS business models in a generic way, I came to the conclusion to build on existing models rather to develop my own classification. Screening various approaches to classify OSS and other business models (e.g., Alexy 2009, Behlendorf 1999, Bonaccorsi et al. 2006, Doganova & Eyquem-Renault 2009, Fitzgerald

2006, Hecker 1999, Krishnamurthy 2005, Lakka, Stamati, Michalakelis & Martakos 2011, Lindman, Rossi & Puustell 2011, Olson 2005, Perr et al. 2010, Riehle 2009, Riehle 2011b, Watson et al. 2008, Wesselius 2008) revealed that the type of license, ownership, and the absence or presence of a developer community are distinguishing features. Regarding the importance of a revenue model, I excluded approaches that do not aim at receiving a revenue stream allocated to OSS. Therefore, in contrast to previous work, I do not consider cost and risk reduction, business transformation, or outsourcing (e.g., Ågerfalk & Fitzgerald 2008, Alexy 2009, Fitzgerald 2006) – although sophisticated OSS strategies – as OSS business models.

Finally, Watson et al. (2008) suggested five distinct models of software production and distribution; three of them may be regarded as business models (a classification I fully support) namely, *Corporate Distribution*, *Sponsored Open Source*, and *Seceond-Generation Open Source*.

**Corporate Distribution.** This model builds on the open community model, in which development involves support of volunteers with limited or no commercial interests (Watson et al. 2008). Many projects with an open community model receive very little attention and consist of a limited number of programmers. Often, these projects are rather small and aim to solve individual user's needs. However, some projects have the potential to be of value for firms as well. Firms that realized that user firms may benefit from such software projects (e.g., RedHat and Suse in the case of Linux) started to build up distribution channels and marketing for these products. Although it was possible to download Linux for free, by the end of the last century, only a few users had Internet connections with adequate bandwidth. Packaging Linux, providing manuals, and selling the software in stores was therefore a reliable business model at that time as the price for the package was still considerably below the price for Microsoft Windows 98. Consequently, various scholars refer to this model as distributor model (e.g., Krishnamurthy 2005).

Soon distributors realized not only could they sell Linux to the average end user but also to companies, especially as an operating system for servers. However, selling OSS to business customers is different from selling to consumers in many ways.[13] For example, although a firm might run several hundreds of servers, in a pure distributor model, the distributor only receives rents for one installation.[14] In addition, OSS was long thought of

---

[13]For an extended discussion pertaining to the difference between Business-to-Business (B2B) and Business-to-Consumers (B2C) see Walsh, Klee & Kilian (2009).

[14]One of my interviewees reported, that 1&1, a major German reseller for Internet connections, bought

as being developed by volunteers and hobbyists only. As this gave rise to the assumption of lower quality, distributors had to spend money on building trust in OSS products. Another concern of user firms was that they are dependent on a running system. However, if indispensible programmers are surfing while their knowledge is needed for solving problems immediately, this causes major threats to business customers.

In turn, distributors began to pay specific developers of the project on which their business was dependent. As we will see in Chapter 5, RedHat still pays the majority of developers for the Linux kernel. Yet, the need for availability of the software was also a new market for distributors. Consequently, they began to offer service and maintenance to their customers, both of which today makes up the majority of their income. Thus, the distributor model falls into the category of selling complementary services.

**Sponsored Open Source.** Many firms sponsor OSS just as local merchants sponsor local unprofessional soccer teams. However, professional soccer teams are usually sponsored by large companies, such as telecommunications companies or insurance companies. In a similar vein, Iansiti & Richards (2006) pointed out that approximately 99% of all sponsorship goes to only 1% of all OSS projects. This seemingly reflects strategic reasons rather than goodwill.

Successful projects are characterized by a high degree of professionalism and are usually represented by a foundation (e.g., Apache foundation, Eclipse foundation, Linux foundation, Mozilla foundation) as a means to limit individual legal liability (O'Mahony 2005). The OSS foundation, which may be considered a legal entity supporting the OSS community, receives donations and handles all billings. Many firms sponsor OSS projects without the intent to control or influence development.

For example, Wagstrom (2009) reported for the Eclipse foundation, whose members are institutions rather than individuals, that even marginally active firms sent checks to pay their membership fee every year, a fee that ranged from $5,000 to $500,000. Obviously, those firms have no interest in specific project trajectories, they only want to keep the software alive. Usually, those firms do not have a business model dedicated to OSS nor are they software vendors.

Another example pertains to the protocol AMQP. Various firms, such as JPMorgan Chase, Cisco, or Deutsche Boerse sponsor a project that aims at providing a protocol for

---

one single Linux distribution from Suse for DM 129. As they possessed the necessary skills, they were able to use this instance to clone their entire server infrastructure. Later, this option was prohibited by distributors.

data exchange. None of the sponsoring firms has a direct financial interest in the product, that is, no firm intends to sell the code or provide complementary service. They engage simply because they expect to save costs. So far, sponsored open source is not a business model.

Admittedly, firms that have an interest in the direction of development do sponsor OSS communities strategically. Moreover, firms with a business model dedicated to OSS assign own developers to the project and provide additional resources. The group of sponsoring firms strategically embraces hardware sellers (e.g., producers of mobile phones or game consoles), software producers (e.g., firms that integrate OSS code into their own code base), and, to a lesser degree, distributors.

In order to distinguish the distributor model from sponsored open source – a task which is difficult in some situations –, one might think of sponsored open source as the sale of complementary products.


**Second Generation Open Source.** According to Watson et al. (2008), a second-generation open source (OSSg2) firm is a hybrid between corporate distribution and sponsored OSS. In essence, hybrid business models are not generic ones that ought to lead to their exclusion from business model classifications. However, in the case of OSSg2, it is worthwhile to make an exception as we will see.

OSSg2 companies generate the lion's share of their revenues by providing complementary services to their products. In addition, they provide the bulk of the resources required to create software products themselves and thereby show similarities to classical software vendors (Riehle 2011b). However, in contrast to the other OSS business models, OSSg2 firms own the IP of the OSS products either entirely or almost entirely. Thus, they are able to tightly control the projects' development. Regarding Dahlander's (2007) framework, the majority of OSSg2 projects were initiated by a single company. Instead of keeping the software closed (i.e., a proprietary approach), the software is offered as OSS from the beginning and a community usually does not exist and must be built. Although someone could argue that OSSg2 is nothing other than a proprietary software vendor accepting the relinquishment of license fees, this is only half the truth.

Firstly, following Watson et al. (2008), no proprietary vendor has the degree of assurance an OSS product has. As anyone can look into the code itself, a mandatory characteristic for OSS licenses, circumstantial programming behavior as well as hidden bugs can be detected *before* the customer firm actually uses the software. Secondly, many of the

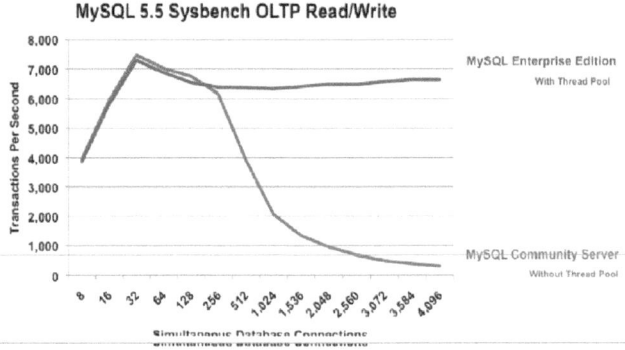

**Figure 3.5** Performance of MySQL: Community vs. Enterprise Edition

OSSg2 firms do sell licenses – but not the open source license. Since OSSg2 firms own the IP to their code, they are able to provide *different* licenses to different groups. This approach is known as *Dual Licencing* and refers to the strategy to provide OSS licenses with which a community can work, and a commercial license that allows for business use. Self-evidently, user firms have to pay for the rights to use the commercial version, which in most cases additionally embraces enlarged functionalities (Olson 2005). Figure 3.5 shows that for MySQL, the enterprise edition is able to handle more requests simultaneously than the free community edition. Sleepycat and JBoss are also prominent examples of OSSg2 companies.

Comparatively early, Krishnamurthy (2005) refered to OSSg2 firms – a term that was not present at that time – as OSS producers with a non-GPL model business model. With GPL, derived work has to be published under GPL again (see Section 3.1.2), so non-GPL licenses allow for integrating OSS code in proprietary code. Thus, whereas GPL code is not very business friendly, non-GPL licenses provide an incentive for user firms to even pay for OSS. Krishnamurthy's (2005) non-GPL model may therefore be considered a prerequisite for a Dual Licensing approach. Figure 3.6 illustrates the basic mechanism of the non-GPL model.

In summary, Watson et al. (2008) provided three OSS approaches that may be considered as OSS business models and that extend the generic notion of selling complementary products and services in the absence of direct revenue streams. Their classification also showed deficiencies, such as the non-consideration of ownership structures as a distinguish-

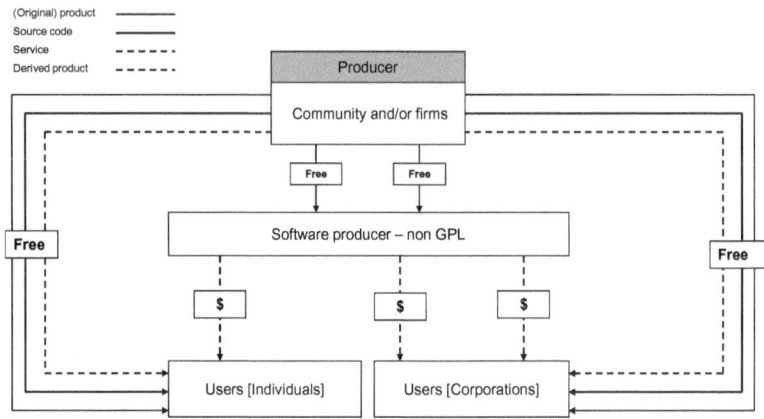

**Figure 3.6** Software Producer – Non-GPL Model (Source: adapted from Krishnamurthy 2005)

ing criterion. For instance, OSSg2 firms can, but do not have to apply a Dual Licensing approach in order to be classified as OSSg2. In contrast, other authors, such as Alexy (2009) or Riehle (2011a), treated Dual Licensing as an OSS business model on its own. However, the delineation of corporate distribution, sponsored OSS, and OSSg2 according to Watson et al. (2008) captured as many important dimensions as possible and is therefore sufficient to act as a basis for this thesis.

### 3.3.4 Hybrid Open Source Business Models and Side Effects

Most classifications of OSS business models stem from real-life examples of OSS firms. Not surprisingly, these firms fit well into their allocated business model. However, apart from well known examples of how to capture value from OSS, such as RedHat, MySQL, or JBoss, it is difficult to categorize de novo OSS firms in terms of their business model. Often, new firms make use of different mechanisms simultaneously, which makes it difficult to classify their approaches. This might explain the evolving variety of business model categorizations. However, the bulk of de novo OSS firms use hybrid approaches. In particular, as noted by Watson et al. (2008), even OSSg2 is a hybrid of corporate distribution and sponsored OSS. Thus, in many cases it would be useful to speak of a firm as applying, let's say, business model type A to 20%, business model type B to 30% and business model type C to 50%. It is self-evident that such kind of classification would not minimize the level of complexity and thus are comparatively useless. Thus, instead of providing different examples of mixed or hybrid OSS business models along with various strategies, I conclude by offering additional (positive) side effects that stem from applying OSS strategies.

Augustin (2008) has pointed in various speeches to the internal financial structure of software firms. In the golden years of software development at the end of the last century, on average software vendors had a balance sheet with 87% gross margins, 17% operating income, 41% of revenue (or 60% of expenses), going to Sales and Marketing (S&M), and 18% of revenue (or 25% of expenses) going to research and development (R&D). Accordingly, the spending for S&M was 2.3 times the spending for R&D on average (Table 3.4).

**Table 3.4** Comparison of the Internal Financial Structure of Software Vendors in the 1990s and After 2000 (Source: Augustin 2008)

|            | 1989-1999        | 2000-2008             |
| ---------- | ---------------- | --------------------- |
| S&M spend  | 2.3 x R&D spend  | 6 to 11 x R&D spend   |
| R&D spend  | 25% of expenses  | 7 to 12 % of expenses |

After the year 2000, however, things began to change. Customer demanded more sophisticated solutions and competition in the software market increased rapidly due to a high number of niche vendors. As maintaining IT environments became more important to software vendors, the revenue profiles changed. As the IT budget of user firms did not increase that rapidly, commercial customers were likely to negotiate the price for the software license. However, software firms needed the revenues from licenses to respond to increasing competition by strengthening their marketing. In particular, the expenses for fairs and exibitions, for salespeople and for advertising campaigns grew dramatically. By 2008, 81 % of new license revenue went to S&M for large enterprise software and the spending for S&M was 6 to 11 times the spending on R&D (Augustin 2008). In other words, instead of building better products, firms invested in convincing the customer that he ought to buy a superior product.

Regarding the dominance of marketing activities for software products, OSS may help to reduce costs for S&M effectively. Since OSS can be downloaded from the vendor's homepage, a potential new customer can download the software, install it, and check if it fits to its IT environment. For this procedure, no salespersons with expensive cars and travel expenses are needed. Augustin (2008) spoke of "turning downloads directly into dollars." In a similar vein, West & O'Mahony (2008) reported that OSS can reduce entry barriers for start-ups, since the barriers for new entrants stem from not having $ 50 million in marketing rather than not having technical experience. In addition, an interviewee of a German OSS marketing company told me that sales cycles for enterprise and business software such as Enterprise Resource Planning (ERP) may take up to two years from the first contact to the final handshake. OSS can seemingly help shorten this sales cycle significantly.

Most OSS firms are privately held and therefore do not have the obligation to publish their figures. However, by referring to RedHat, which had its IPO in 1999, Augustin (2008) argued that OSS firms pretty much look like pre-bubble burst software vendors. In particular, by 2007 RedHat had 83 % Gross Margin, 25 % of expenses going to R&D, and 55 % of expenses to S&M (see Table 3.5).

**Table 3.5** Comparison of the Internal Financial Structure of Proprietary Software Vendors and OSS Firms (Source: Augustin 2008)

|               | RedHat           | Traditional Model |
|---------------|------------------|-------------------|
| Gross Margins | 83%              | 87%               |
| R&D           | 25% of expenses  | 25% of expenses   |
| S&M           | 55% of expenses  | 60% of expenses   |
| G&A           | 20% of expenses  | 15% of expenses   |

In summary, beyond the discussion on OSS business models, the OSS development approach challenges traditional ways of software development and distribution. In particular, OSS can speed up a rapid diffusion of the product due to the absence of an up-front payment, generate perpetual revenue streams through complementary services, and has the potential to reduce S&M expenses. The challenges for firms, however, still stem from the fact that different stakeholders in an OSSDP might have different interests that lead to tensions. The following section is therefore devoted to governance and control in different forms of OSS development.

# 3.4 Extending Control Theory for Managing Open Source Innovation

As pointed out, several ways exist for firms to benefit from engagement in OSS. No matter which form of OSS engagement is chosen, a community of different stakeholders surrounding an OSSDP exists per se. However, the constitution of the community fully depends on the type of OSS business model (Bonaccorsi & Rossi 2006). For example, newly founded SVPs do not have a community with voluntary external developers. The community is mainly built by developers paid by the single firm and contingently built by users. In contrast, in large OSS communities such as Apache or Linux, many paid developers of different firms and volunteers together build what is called the community. Since firms tend to work with the community of voluntary developers in order to increase their inventive and absorptive capacity, recent research has highlighted the importance of community management (O'Mahony 2007). Accordingly, firms that are dependent on a community consisting of other paid developers and volunteers as a pool of complementary resources (Walsh et al. 2011) must know how they communicate and interact with their community.

The following sections therefore embrace the role of the community as a boundary organization and explore different tensions that arise by means of different interests and governance in organizations. Building on control theory, which distinguishes formal from informal control mechanisms (Aghion & Tirole 1997, Aiken & Hage 1968, Carver & Scheier 1981, Kirsch 1997, Ouchi 1979, Ouchi 1980, Rustagi, King & Kirsch 2008, Snell 1992), I pursue a better understanding of different control mechanisms in the context of OSS communities as complementary resources and thereby extend established models for governance and control.

## 3.4.1 Boundary Organizations for Open Source Innovation

Grand et al. (2004) used the term *open source innovation* to refer to the specific challenges that arise with considering OSS communities as complementary resources in an open innovation sense. For example, firms that engage in the development of OSS have obligations to the community that supports the OSSDP (Ågerfalk & Fitzgerald 2008). These obligations require accepting the community's norms and beliefs as well as aligning with its goals. If the firm intends to influence the community by shifting norms, beliefs and goals of the community to a set of activities that are consistent with those of the firm, community management is needed (O'Mahony 2003). O'Mahony (2007, p. 141) argued that "management has become decoupled from the notion of open source and that it is deserving of further inquiry in and of itself – particularly in light of the hybrid open source communities that are emerging." However, as many others have, she has pursued a firm's perspective by regarding the community as something outside the firm, a notion that is inherent in the term *community management*.

Another and more abstract perspective was offered in a follow-up work by O'Mahony & Bechky (2008). In particular, they treated OSS communities as *boundary organizations* that negotiate between actors with different interests. Like individual boundary spanners, which are active in different domains and therefore possess knowledge in both domains to bargaining between different actors (Tushman & Scanlan 1981b), boundary organizations mediate different parties that have begun to ally.

Regarding boundary organizations, in accordance with O'Mahony & Bechky (2008), boundaries are considered the result of social movements. In order to provide an example, they referred to the restructuring of the recycling industry. Unlike at the beginning, when recycling was considered a marginal practice promoted by activities, today, recycling is a profitable business. In a similar vein, the current social movements such as the anti-nuclear power movement in Germany lead to restructurization of the energy industry which, in turn, impacted the firm boundaries of energy companies. Thus, from a theoretical point of view, O'Mahony & Bechky (2008) called for managing boundaries in collaborations rather than managing communities.

By following Snow, Soule & Kriesi (2004), O'Mahony & Bechky (2008) distinguished challengers from defenders of existing institutional authority. Often, challengers are likely to constitute a movement as a form of protest. In this sense, OSS activists have long been considered promotors of protest against multi-national software companies, such as Microsoft. However, as we have seen in section 3.1, people engaged in the development of OSS are driven by a portfolio of motives, rather than exclusivly driven by a protest attitude. Consequently, OSS communities can also be seen as boundary organizations.

As in other forms of organizations, a community will only be formed if costs of organizing the group will be smaller than the benefits the group gains (Weber 2005). In a similar vein, firms will only tend to change the composition of the OSS community their business model is dedicated to if the costs for restructurization do not outperform the benefits. This behavior can also be explained by transacton cost economics since assigning their own developers to work for an OSSDP, an action that affects the composition and structure of a community, may be considered an expansion of firm boundaries (cf. Demil & Lecocq 2006, Williamson 1991).

In the case of OSS, it is not that easy to identify the defenders and challengers of the status quo. For example, if voluntary programmers contribute to OSS as a form of protest against proprietary software vendors, they make themselves challengers. On the other hand, if firms intend to cannibalize OSS communities and to change their structure, firms may be considered challengers of the current system. However, what both scenarios have in common is that (at least) two different parties exist that share a certain goal but show differences in their interests, attitudes, and beliefs.

### 3.4.2 Tensions of Convergent and Divergent Interests in Boundary Organizations and the Emerging Need for Control

O'Mahony & Bechky (2008, p. 425) felicitously argued: "if challenging and defending parties are to ally, they must find a way to bridge their differences without threatening the core values that make them distinct." As pointed out in the previous sections and made explicit by Dahlander (2007, p. 915), "the incentive structures underlying innovation in private firms and innovation in commons are often thought to be fundamentally different." For example, firms traditionally have an interest in commercializing their inventions and try to protect their inventions from misuse by IPRs but in some situations, protection is difficult to obtain. However, once in the public domain, information cannot be privatized. As noted by Merges (2004), if it is in the firm's interest to preempt an asset from being privatized, the firm will invest in the creation of that asset and then inject it into the public domain. By applying this strategy, firms contribute to the public domain while pursuing private interests.

Consequently, given that actors possess multiple interests in different contexts, firms and the commons (i.e., the group of voluntary developers in OSS communities) do have overlapping interests. For instance, West & O'Mahony (2008) argue that attracting greater growth of participants is an interest that firms and the community of volunteers share. However, at the same time they have fundamentally different interests due to their specific beliefs.

I refer to the latter group of interests as divergent interests while using convergent interests to refer to the former group of interests.[15] Table 3.6 summarizes overlapping and contradictory interests.

**Table 3.6** Convergent and Divergent Interests of Firms and OSSDPs (Source: O'Mahony & Bechky 2008, p. 432)

| Community-managed OSSDP | Firms |
|---|---|
| *Convergent interests* | |
| Enhance technical capability, performance, and portability of software for use in the enterprise | Acquire access to technical expertise and improve recruitment of skilled programmers |
| Improve individual skill through exposure to new commercial performance | Collaborate with skilled experts to solve difficult technical problems; learn how source code can be customized to solve customer problems |
| Achieve commercial legitimacy and recognition; establish traditional marketing challenges | Alleviate power of industry monopoly and enhance their own market share |
| Enhance project's market share and diffusion | Increase margins through reduced licensing fees |
| *Divergent interests* | |
| Maintain communal form: informal collegial project practices and working norms | Influence project direction to align with firm strategy and timetable |
| Maintain individual technical autonomy | Acquire more predictability in the software development process to foster firm planning |
| Preserve transparency and open access to code development in order to foster full participation in community decision making | Pursue partnership and collaboration with descretion |
| Sustain project's vendor independence | Establish formal governance mechanisms to shape a project's future |

Regarding convergent interests, both firms and the community-managed OSSDP benefit from enhanced technical capability of the software: Community members because they fulfill their motivation to participate, firms because they can access improved technical

---

[15]By using the term *community-managed*, O'Mahony & Bechky (2008) excluded SVPs from their arguments. In order to stay coherent with my wording, I prefer to use OSSDP at this point, knowing, that OSSDP embraces SVPs as well.

expertise. Developing high-quality software and solving difficult technical problems thus are connected interests. Similarly, enhancing the project's market share and diffusion suit the community and the firm. However, while the firms' interest in influencing the OSS development tends to increase with their investments, the community does not want to have the product aligned with specific firm strategies. Instead, as in other cases of virtual collaboration (e.g., Davenport & Daellenbach 2011), the community wants collegial project practices and working norms.

A real-life example portrays this divergence. Consider the following situation. A Red-Hat customer's infrastructure runs on Linux. With a new release, the print server does not work anymore. Consider further the old print server only worked because an older Linux version had a bug. How would you decide as RedHat? Do you want to keep the bug in the Linux system to please your customer? Or would you force your customer to invest in a new print server? It's a tough decision, as the customer won't be happy to pay for an already existing infrastructure and the community won't be happy keeping a bug in the next release.

Moreover, if the company decides to promote the bug solution within the community, this decision would not be executed as within a company. Within the boundaries of the firm, employees would be advised to develop in accordance with management's strategy. In contrast, people freely join OSSDPs just because authoritive control is limited (Hars & Ou 2002, Hertel et al. 2003). Thus, motives of voluntary developers and the organization of an OSS community impede the direct execution of firm interests.

From an organization theory perspective, traditional software vendors especially do not favor an open development approach with limited authority. Instead, they prefer hierarchical, controlled, and stable development approaches. This does not mean that OSS communities do not have hierarchical structures. Quite the contrary. OSS communities are led by project leaders and favor a semi-hierarchical structure (Crowston & Howison 2006). However, the leading positions are assigned according to meritocratic performance (cf. Young 1994). As pointed out earlier, meritocracy in the case of OSS refers to the fact that the more you contribute, the more you are allowed to do. Over time, it is possible to climb the community career ladder and capture leading positions. Therefore, although the firm might have its own developers assigned to the OSSDP, it will take awhile until you have proved your programming qualification.

O'Mahony & Bechky (2008) argued that OSS communities act as boundary organizations that negotiate between groups with (partly) different interests. In particular, they define different roles the boundary organization takes on in terms of governance, membership, ownership, and control of production. Regarding governance, a term used in different settings (Preece 2000), the boundary organization ensures independent and collective control without firm control (i.e., the voluntary developers' interest) and provides the firm some voice on the project's trajectory (i.e, the firm's interest). No interest is fully satisfied, but both parties benefit from the boundary organization's role. The same logic holds true for membership, ownership, and control of production (Table 3.7).

**Table 3.7** Role of a Boundary Organization in Enabling Collaboration (Source: O'Mahony & Bechky 2008, p. 441)

| Organizing practices adapted | Interests satisfied | |
| --- | --- | --- |
| | Community-managed OSSDPs | Firms |
| **Governance** | | |
| Establishing project representation | Provides access and participatory processes | Reduces ambiguity and provides some degree of discretion |
| Pluralistic control | Ensures independent and collective control without undue firm influences | Provides some voice on project direction without direct control |
| **Membership** | | |
| Defining rights of members | Preserves individual basis of membership and independence of the community | Firms cannot gain formal rights, only sponsor contributors |
| Sponsoring contributors | Provides additional resources to help project improve | Offers firms a means of direct access to development process |
| **Ownership** | | |
| Obtaining work assignment rights | Reinforces individual autonomy and independence | Ensures clear provenance of code |
| Developing contribution agreements | Ensures clear provenance of code | Ensures clear provenance of code |
| Managing code donation | Enhances technical quality and reach of the project | Improves efficiency: no separate code base to manage |
| **Control of production** | | |
| Community control of code contribution | Allows community to preserve autonomy and independence | Sponsored contributors provide firms with visibility and access to code development |
| Managing technical direction | Allows community to preserve autonomy and independence | Sponsored contributors provide firms with informal influence on code development |

In summary, the boundary organization is a means to satisfy both the firm's and the community-managed OSSDP's interests where they are different. However, as discussed by O'Mahony & Bechky (2008), the control and influence of firms in the technical development remains informal. Thus, in the following I offer alternative ways of influencing and controlling OSSDPs based on an understanding organizational control according to Ouchi (1979).

### 3.4.3 Governance and Control in Organizations

Most governance literature stems from two streams of research, from political science and corporate management (Ruhanen, Scott, Ritchie & Tkaczynski 2010). The research sparked by corporate management ideas tends to consider questions of how organizations can exert control over strategically important activities and processes on a micro level (Choudhury & Sabherwal 2003, Kirsch 1997, Ouchi 1979) and how to interact with external parties on a macro level (Eisenhardt 1985, Ouchi 1977, Williamson 1985). In line with transaction costs economics (Williamson 1985), the governance modes on a macro level represent forms of cooperation, ranked along a continuum from integrated (or vertical integration) to market transactions (Van de Vrande, Vanhaverbeke & Duysters 2009).

In organization theory, governance on the micro level combines different mechanisms to encourage people to do things that align with the organization's preferences (Choudhury & Sabherwal 2003), so it can be defined as "any process by which managers direct attention, motivate, and encourage members to act in desired ways to meet the firm's objectives" (Cardinal 2001, p. 22). Because control entails a portfolio of mechanisms designed to influence employees (Cardinal 2001, Elias 2009, Jaworski 1988, Kirsch 1997), governance and control are often and confusingly used interchangeably.

Instead, I specify a behavioral view of control, defined as "attempting to ensure individuals [...] act in a manner that is consistent with achieving desired objectives" (Choudhury & Sabherwal 2003, p. 292). I draw on the broader framework developed by Ouchi (1977, 1979), which is based on antecedent conditions such as the level of task programmability (i.e., knowledge about the transformation process available to a controller) and output measurability. Depending on whether the task is definable and the outcome of the task is measurable, behavior, outcome, and clan control are distinguished.

For example, output control is suitable only if it is possible to measure employee activities, such as counting output produced by a factory worker (Cardinal, Sitkin & Long 2004, Eisenhardt 1985). Behavior control instead seeks to influence employees by defining rules and procedures to be followed (Kirsch 1997, Ouchi & Maguire 1975). Outcome control requires relatively little management direction, but behavior control depends completely on monitoring systems, which then require hierarchical management layers. Anderson & Oliver (1987, p. 77) analyzed control systems for salespeople and argued that in behavioral control settings, managers must have a "well-defined idea of what they want salespeople to do and work to ensure the salesforce behaves accordingly."

However, in the absence of clear task specification and outcome measurability, social control is more likely within organizations; it can be achieved by minimizing divergent preferences among organizational members through socialization (Eisenhardt 1985). By internalizing a company's goals, every employee shares the company's vision and contributes to the organizational culture. Control is particularly important in cases of divergence, and common goals, norms, and beliefs help decrease the level of required control, in that the group pursues ceremonies and acts like a clan (Alvesson & Robertson 2006, Kirsch, Ko, & Haney 2010, Ouchi 1979). Such clan controls, based on shared norms and beliefs, are particularly observable in OSS communities (Stewart & Gosain 2006) and start-ups (Stevenson & Jarillo 1990) that eliminate the applicability of other forms of control because of they lack formal structures (Lattemann & Stieglitz 2005). Figure 3.7 provides an overview of the different control modes according to Ouchi (1979).

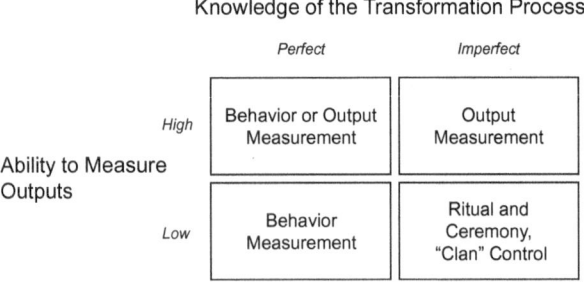

**Figure 3.7** Conditions Determining the Measurement of Behavior and of Output (Source: Ouchi 1979, p. 843)

Whereas behavior and outcome controls constitute formal mechanisms embedded in a bureaucratic organization design (Rustagi et al. 2010), clan control is informal (e.g., Aghion & Tirole 1997, Blau & Scott 1962, Ouchi 1977, Kohli & Kettinger 2004). A behavioral view of control also implies that each control structure is based on different mechanisms to promote employee behavior; a mechanism that works for one type of control does not work for another (Kirsch 1997, Ouchi & Maguire 1975). However, an often neglected influence mechanism over employee behavior is leadership, which is scarcely discussed in prior research pertaining to control (e.g., Kirsch 1997, Ouchi 1979). Similar to control, leadership comprises formal leadership, achieved through positions defined by a job description, and informal leadership, which likely results from the leader's charisma and has little to do with his or her position in the organization (Conger & Kanungo 1987). Leadership in a bureaucratic sense implies decision-making competence, which may be the highest form of control (e.g., Arya, Glover & Sivaramakrishnan 1997). Thus leadership offers strong access to control in organizational settings (e.g., Choudhury & Sabherwal 2003, Druskat & Wheeler 2003).

In summary, control theory offers a framework to illustrate how environmental factors and control mechanisms depend on different antecedents and influence organizational outcomes variously. Control structures can be classified into outcome and clan control (Ouchi 1979), each of which relies on different mechanisms, such as hierarchy, job description, defined tasks, or implemented rituals and ceremonies (Kirsch 1997). Leadership, both formal and informal, can constitute a strong path to control in each control mode. However, control structures within an (bureaucratic) organization rely on threatening consequences, so it becomes difficult to implement them in systems based (at least partly) on volunteer work (Lattemann & Stieglitz 2005); these settings instead require trust and intrinsic motivation as more adequate governance mechanisms (Shah 2006, Stewart & Gosain 2006). I therefore need to investigate how firms might seek to obtain control in OSSDPs.

### 3.4.4   Firm Control in Boundary Organizations

Generally, a firm's main function should not be to control employees. Control is costly, because it requires monitoring activities (Anderson & Oliver 1987), may cause negative effects when perceived by employees (Brockner, Spreitzer, Mishra, Hochwarter, Pepper & Weinberg 2004), and thereby hinders innovation (Cardinal 2001).

However, need for control increases with organization size and task divergence (Ouchi 1977) because as an organization grows, more ideas and visions arise that must be aligned to achieve the firm's initial goals. In addition, firms that seek to benefit from an enlarged knowledge base by integrating external resources experience greater pressure to keep the project on track, because the divergence represented by the participants increases (Almirall & Casadesus-Masanell 2010, Grand et al. 2004, Piccoli & Ives 2003, Sundaramurthy &

Lewis 2003). Accordingly, West & O'Mahony (2008, p. 152) note that a primary challenge that firms face is "how to manage the tension between controlling the community in order to leverage the firm's investment in it and opening up access to the community in order to attract greater growth of participants."

Considering the absence of contract-based regulations and intellectual property protection in OSSDP, perceived divergence might be even higher because the firm's own workers, employees of other firms, and voluntary contributors all try to influence the project trajectory (Bonaccorsi & Rossi 2006). Governance then should vary for paid human resources versus freely available ones. A paid employee, generally speaking, accepts certain responsibilities in exchange for remuneration, such as following a supervisor's instructions. In contrast, voluntary contributors are not contractually linked to the firm and may choose to retire from the project at any time (Shah 2006). Relying on contributions from volunteers may thus be disadvantageous if the firm cannot bind these key contributors.

A community of diverse stakeholders limits the application of what Amin and Cohendet (2004) called "management by design," namely, management based on hierarchy, orders, and blueprints, because firms need to acknowledge the culture of the OSSDP. Any OSSDP, initiated by a firm or a community, features unwritten norms and beliefs (Bergquist & Ljungberg 2001, Stewart & Gosain 2006, Von Hippel & Von Krogh 2003). For example, a meritocracy gets implemented in nearly every OSSDP such that the more a developer contributes and the higher the quality of his or her contribution, the more that person may contribute to core functionalities or release cycles. According to recent research, leadership positions in an OSSDP correlate with sustained contributions in advance (Dahlander & O'Mahony 2011, Scozzi, Crowston, Eseryel & Li 2008).

In line with the control modes proposed by control theory (e.g., Anderson & Oliver 1987, Choudhury & Sabherwal 2003, Kirsch 1997, McMahon & Ivancevich 1976, Ouchi 1977, 1979) I argue that firms that want to influence a project's trajectory have two options. The first depends on the firm's ability to install its own employees as project leaders, perhaps by giving those employees sufficient time to climb the meritocracy ladder in the community (Dahlander & O'Mahony 2011, Giuri, Rullani & Torrisi 2008) or by hiring existing project leaders. When employees appear in leadership positions and possess decision-making responsibilities or license or organizational sponsorship choices (Giuri et al. 2008, Heckmann, Crowston, Li, Allen, Eseryel, Howison & Wei 2006, Henkel 2009, Stewart et al. 2006), the firm can influence and control the OSSDP as it desires, because the firm controls the project leaders outside the OSSDP. To avoid bearing ill consequences in cases of bad performance or misbehavior, a project leader likely acts in accordance with the employing firm's interests. I refer to this form as control by leadership.

Formal leadership relates closely to formal control modes, which demand outcome measurability and knowledge about the transformation process (Ouchi 1979). Despite the potential to track developers' behavior by monitoring their output over concurrent versioning systems, the ability to judge outcomes on an individual level remains difficult, because the software code developed by a group cannot be easily traced to the individual programmer's contribution (Stewart & Gosain 2006). In addition, as in any innovation process, knowledge about the transition from input to output is limited, which violates important prerequisites for the application of formal control modes (Eisenhardt 1985).

**Table 3.8** Mechanisms of Control for Firms Participating in OSSDP

|  | Behavior control | Outcome control | Clan control |
|---|---|---|---|
| Antecedent condition | Knowledge of appropriate behaviors | Outcome measurability | Appropriate behaviors unknown |
|  | Behavior observability |  | Outcomes not measurable |
| Mechanism in firm | Job description | Defined target | Socialization |
|  | Hierarchy | Expected level of performance | Rituals and ceremonies |
| Mechanism in community | Leadership | Leadership | Socialization |
|  | Project milestones | Project milestones | Rituals and ceremonies |
| Mechanism firm in community | Acquisition of project leaders | Acquisition of project leaders | Assigning developers socialized within the firm |
|  | Control by leadership | Control by leadership | Resource deployment-based control |

Therefore, firms might apply the second control option which I call resource deployment-based control. In the absence of prerequisites for formal control, social or clan control that reflects shared goals and ideology is appropriate (Barker 1993, Stewart & Gosain 2006). For OSSDP, I must distinguish between two clans – one within the community and one within the focal company. As an employee of the firm and a member of the OSSDP community, a developer becomes embedded in two different systems of norms and beliefs (Dahlander & Wallin 2006, Henkel 2009). By assigning programmers who have internalized firm norms to an OSSDP, the firms might leverage their own resources to obtain control over the project, without formally applying for leadership positions. This resource deployment-based control approach differs from classical clan control, where shared goals and aims evolve within the community. However, injecting firm-socialized resources into a community increases the likelihood that the community norms and values evolve in a way congruent with those of the firm. In Table 3.8 (see also Kirsch 1997, p. 219 for a comparison), I provide an overview of different control mechanisms that firms can apply to achieve control over an OSSDP.

## 3.5   Summary

The aim of this chapter was to explore the OSS phenomenon as a form of innovation beyond firm boundaries. Moreover, various approaches to create and capture value with OSS were delineated, such as the sale of complementary products and services. Screening various classifications of OSS business models lead to the acceptance of Watson et al.'s (2008) scheme that delineates corporate distribution from sponsored open source and OSSg2.

In cases of SVPs, a single firm usually owns the entire rights to the software. In this situation, controlling technical development is more or less equal to proprietary approaches. The way how marketing is pursued is what is different from traditional software vendors. However, in cases of distributed ownership, controlling the project's trajectory is by far more difficult since multiple groups aim at satisfying their interests. Building on the work of O'Mahony & Bechky (2008), I further explained the role of a boundary organization as a mediating entity.

Starting from the assumption that firms cannot apply direct and formal control over a community-managed OSSDP, I suggested alternative ways for firms to control the development, such as what I call resource-deployment based control. Instead of waiting until paid programmers have reached leading positions within the community, a firm can influence development trajectories by assigning many developers that have been socialized within the focal firm. Thereby, the firm can apply a form of clan control within the community and is capable of managing innovation beyond firm boundaries.

# Chapter 4

# Open Source in Action I: Business Collaboration Among Open Source Projects

## 4.1 Introduction

It is generally assumed that in times of high competition, the ability to innovate is of high importance to corporate R&D. Moreover, a growing body of management research takes as its central assumption, that opening the boundaries of the firm potentially increases its innovative performance (Chesbrough 2003b, Laursen & Salter 2006, Lichtenthaler 2010a, Enkel, Gassmann & Chesbrough 2009). In addition to the general trend of opening R&D processes to external participants (Chesbrough 2006), which is currently discussed under the term open innovation, a further notable (but not necessarily new) trend is that firms increasingly make use of R&D alliances and consortia (Garringa, Von Krogh & Spaeth 2011).

With the emergence of OSS as a development approach, alternative ways of organizing R&D consortia appeared as its characteristics enable programs to coalesce as a result of a collective work of hundreds of independent programmers distributed over the world. As outlined previously, over the last few years, the development of OSS has changed due to the participation of firms in existing OSS projects and the firm-driven initiation of new projects, which usually start with a firm's codebase (Dahlander & Magnusson 2008). Although the phenomenon of firm participation in OSS projects is not entirely investigated, recent studies revealed a number of economic and strategic reasons (Alexy 2009, Fitzgerald 2006, Fosfuri et al. 2008) (see Section 3.3.3). First, in the sense of open innovation, working with a community of well-experienced programmers enlarges a firm's social capital and innovative performance (Roberts et al. 2006). Second, using contributions from external individuals is predicted to save costs (Lerner & Tirole 2002). And finally, if a program no longer contributes to the firm's value creation, such as in the case of Netscape's Navigator, giving the code to a number of volunteers for free can prevent competitors from harvesting a market (Fershtman & Gandal 2007, West & Gallagher 2006).

As in other forms of collaboration, the coexistence of a profit-oriented firm and a community of voluntarily engaged programmers in one project inevitably leads to tensions. To comply with firm interests, firm-initiated OSSDPs especially are predicted to be led by project leaders sponsored by firms with an interest in the project. (Dahlander & Wallin 2006, Henkel 2009). In contrast, self-realization and freedom are one of the most prominent motivations for individuals to participate in OSS development (Bitzer et al. 2007, Shah 2006). However, enlarged firm presence in OSSDPs may drive voluntary developers to deallocate their resources from that project (Walsh et al. 2011). In addition, the development styles, namely a bazaar style in the case of OSS and a more structured one run by software vendors, are not compatible (Raymond 1998). Thus, as a result of different interest (cf. Section 3.4), those projects contain a number of natural tensions like the one between collaboration (of a diverse set of voluntary developers and a firm) and (firm) control (Sundaramurthy & Lewis 2003, O'Mahony & Bechky 2008).

Regarding R&D consortia, which are multi-vendor projects (MVP) by definition in the case of OSS development, West and Gallagher (2006) point to pooled R&D, a mechanism to cooperate in research using the OSS model not only to save costs but also to address industries with strong vertical relationships or situations where firms cannot appropriate spillovers from their investment (Sakakibara 2002). In the case of OSS, so-called foundations have been created by contributors to function as a legal umbrella for OSS projects. These non-profit organizations aim to minimize possible negotiations between different participants, provide projects with a set of verified (and business friendly) licenses, and help promote the OSSDP (O'Mahony 2005).

As such, foundations may take over the role of boundary organizations and dissolve some problems associated with different interest in OSS development, but at the cost of new dilemmas. For example, without a foundation there is just a community of contributors; no matter if paid (i.e., firm-affiliated) or unpaid (i.e., voluntary). However, the threat inherent in installing a foundation is twofold. Without a foundation, investors possibly would withdraw their engagement due to missing professionalism. With a foundation, the OSSDP risks to lose its OSS identity. Creating a foundation, thus, requires to communicate to several stakeholders (i.e., press, users, investors) where to draw the line between OSSDP and the firm. The majority of foundations, such as the Apache Foundation, have been created to promote a single OSSDP. However, with the increasing commercialization of OSSDPs, foundations have begun to host several projects. For instance, the Mozilla Foundation takes over responsibility for Firefox, Thunderbird, SeaMonkey, Camino, and several other projects.[1]

Although quite a number of studies have dealt with the coexistence of firms and a community in OSS settings (e.g., Lee & Cole 2003, O'Mahony & Ferraro 2007, West & Lakhani 2008), the focus of these studies mainly rest on a relationship between a single firm and a community of committers (e.g., Bagozzi & Dholakia 2006, Dahlander & Magnusson 2008). Consequently, investigations regarding the role of different software vendors in one and the same OSS project are underrepresented (Grewal, Lilien & Mallapragada 2006, Méndez-Durón & Garcia 2009). Moreover, there is a dearth of research investigating different forms of OSSDPs simultaneously, such as SVPs and MVPs.

Building upon this line of literature, this chapter complements the dissertation's topic on firms in the development of OSS and their attempts to control OSSDPs by investigating different forms of open innovation and control in boundary organizations, such as OSS foundations (see Chapter 3). First, I will discuss the usage of the OSS development paradigm as an alternative way to establish R&D consortia between more than one firm and a community of voluntary programmers. An empirical investigation will reveal firm behavior consistent with behavior observed in industrial R&D consortia, which is why it is adequate to speak of MVPs to constitute interfirm collaboration. Second, I will analyze different forms of control, such as control by leadership (CBL) and resource deployment-based control (RDBC) as defined in the preceeding chapter, in the context of different forms of OSSDPs, such as SVPs and MVPs.

To address the identified research gaps, I focus on the Eclipse Foundation, which is a member-supported non-profit organization that hosts the Eclipse projects and helps to cultivate both an OSS community and an ecosystem of complementary products and services (O'Mahony, Cela Diaz & Mamas 2005, Wagstrom 2009). I derived a dataset of 912 committers working for 110 different organizations and organized in 109 projects from the Eclipse website that will act as the basis for my analysis.

---

[1]see Mozilla Foundation. http://www.mozilla.org/projects/, last access 8/1/2011

## 4.2 Open Source Software Projects as R&D Alliances

Firms acquire external knowledge in various forms and along different stages in the innovation process like doing market research, cooperating with suppliers or building R&D alliances (Sampson 2005, Simonin 2004). Firms are not only driven by the wish to save costs, but they are searching for competencies outside their own boundaries (Singh 2008). In recent years, many firms opened several parts of their innovation processes for external participation.

According to the different types of possible external sources (users, customers, universities, research centres, competitors, etc.), the degree of integration differs. Whereas users typically do not own the necessary knowledge to contribute to high-technology projects, the contribution from competing firms are valuable for an organization. Most activities integrating users or customers in the innovation process can be observed in early and late phases of the innovation process – like making use of lead users or mass customizing products (Piller & Walcher 2006) – whereas integrating external partners in core development activities is open to a lesser extent (Schaarschmidt & Kilian 2011). In the latter case, firms mainly make use of industrial R&D consortia and joint ventures with other firms that generally require formal agreements (De Rond & Bouchikhi 2004, Sakakibara 2002, Simonin 2004) – without integrating further users or customers.

As discussed detailed in Section 2.2.2, both the integration of external knowledge as well as the external exploitation of intellectual property (IP) is currently discussed under the term open innovation (Enkel et al. 2009, Lichtenthaler 2009b, Lichtenthaler & Lichtenthaler 2009). In contrast to a closed innovation paradigm, firms try to include customers, users, universities, and even competitors in different stages of their new product development processes (Chesbrough 2003c, Lichtenthaler & Ernst 2009a). Furthermore, it is also commonplace to commercialize technological innovation outside the organizational boundaries (Lichtenthaler & Ernst 2007, Lichtenthaler 2009a). If the technology a firm is offering no longer contributes to the firm's value creation, it usually has to be spun out. For example, Chesbrough (2003a) analyzed many technological projects at Xerox and found that some of them were spun out due to missing commercialization options under Xerox leadership. With regard to the different options offered by the open innovation paradigm, Enkel et al. (2009) differentiate between three types of open innovation processes, namely (1) outside-in, (2) inside-out, and (3) coupled (see also Section 2.2.2).

Looking at the OSS paradigm as a special type of open innovation (West & Lakhani 2008), any of the three types are observable – the integration of external contributions to a firm-driven project (1), the publication of software generated inside the boundaries of a firm (2), and both at the same time (3). For instance, Deutsche Post, a major German logistics company, had developed a service-oriented IT infrastructure over the last ten years. Service-oriented architectures (SOA) are a heavily discussed topic in IT settings due to the ability to create a flexible IT infrastructure (Newcomer & Lomow 2005). Recent developments at Deutsche Post made it necessary to offer the in-house solution, an SOA-framework under the umbrella of the Eclipse Foundation, as an openly available product; with hopes for external contributions to the ongoing process of maintaining the software.

Similar arguments can be found by looking at reasons to establish R&D alliances. Beside cost reduction and increased innovative performance, literature mentions reduced risk of technological development, scale economies, reduced time to market, and network effects as the most dominant reasons to build R&D alliances (Arranz & de Arroyabe 2008, Mowery, Oxley & Silverman 1996, Patrakasol & Olson 2007, Sakakibara 2002, Sampson 2005). In the case of OSS, the wish to develop open standards to secure interfirm collaboration on a technical level is an additional factor. As R&D alliances require strong effort in identifying and acquiring possible partners as well as in dealing with legal issues (Vanhaverbeke, Duysters & Noorderhaven 2002), building R&D consortia is often too costly, especially for small and medium sized enterprises (SME), which usually lack resources. For firms, OSS development is therefore a promising way to simultaneously benefit from external knowledge in the sense of R&D consortia – both from knowledge of voluntary committers as well as from competing firms. However, we have to keep in mind, that these approaches inherit the tension of maximizing the created value while at the same time minimizing the threat of opportunism (Abdallah & Wadhwa 2009, Carson, Madhok, Varman & John 2003).

Recent research shows that firms not only cooperate to save costs but also in cases of strong vertical relationships and high technological uncertainty (Sakakibara 2002, Van de Vrande, Lemmens & Vanhaverbeke 2006). Furthermore, it is worth noting that competing software vendors in the same OSS project are often found in infrastructure or interoperability projects. From an economic perspective, firms using the OSS model share resources to develop infrastructure software that they otherwise had to develop on their own. As the output was never intended to generate revenue and oftentimes impossible to commercialize, competing firms do not perceive direct competition and the likelihood of opportunism decreases. In other words, although firms invest resources in such projects, with the contribution of other firms, the relative amount of the investment decreases compared to doing it alone.

## 4.3 Open Source Software and Different Forms of Collaboration

In the preceeding section, I discussed OSS development as a means to collaborate between different firms in the sense of R&D alliances. However, given the amount of possible commercialization strategies and OSS business models, other forms of OSS collaboration exist. This section therefore aims to provide an extended classification of OSS approaches in order to discuss them against different forms of firm control.

I classify OSSDPs as projects with one participating firm, or SVPs, and those in which more than one firm is active, or MVPs. The former are similar to a situation with proprietary software vendors, especially if they feature a dual licensing approach (Augustin 2008, Olson 2005, Riehle 2011b). As discussed previously, the latter tend instead to mimic R&D alliances or joint ventures and might not entail a direct revenue stream. Instead, the multiple firms combine their resources to build a platform and promote certain standards, which will enable them to sell on-top applications and complementary products or services. Moreover, MVPs usually aim to reduce product costs through cooperation in product development.

Control and ownership structures are the key characteristics for differentiating SVP and MVP. Whereas MVPs are controlled by a community of stakeholders (including multiple individual programmers and/or firms), SVPs are controlled by one stakeholder with an explicitly commercial purpose (Riehle 2011b). Consequently, the core business functions differ, including community management, sales, marketing, product management, engineering, and support (Watson, Boudreau, York, Greiner & Wynn 2008). In addition, Dahlander (2007) distinguishes OSSDPs by their impetus, because a project initiated by a firm versus a community likely exhibits different norms and beliefs (see Section 3.3.1). I therefore propose a classification based on these two distinctions, namely the project's control structure and the project's history (Table 4.1).

Each approach is characterized by the number of firms participating and whether it was initiated by a firm or the community of volunteers. MySQL, the famous OSS database, was initiated by a firm that until today owns the entire rights to the code. In contrast, Android was also initiated by a firm, namely Google, but was designed from the beginning as a

**Table 4.1** Alternative Typology of Commercialization Approaches

|                        | SVP                    | MVP          |
|------------------------|------------------------|--------------|
| Firm initiated project | Approach I             | Approach II  |
|                        | JBoss, MySQL           | Android      |
| Community initiated    | Approach III           | Approach IV  |
| project                | SugarCRM, Sleepycat    | Linux        |

project that should attract followers (i.e. other firms). Linux, another MVP example was founded instead by individuals, without firm support in the beginning. Finally, SugarCRM is an example of an OSSDP whose codebase is owned by a single firm, that, however, did not exist at the time the project was initiated.

It is important to note that despite the presence of firms in OSS development, even firm-imitiated projects rely on a committed voluntary user and/or developer base for several reasons. First, without incorporating the community of users and developers, the firm runs the risk of gaining a negative reputation as a free rider. Second, the community offers a valuable resource base for OSS companies, as complementary assets and a pool for potential future employees (Dahlander & Wallin 2006). For example, JBoss recruits its employees almost entirely from the community of voluntary programmers (Watson et al. 2008).

However, in SVPs initiated by a firm, the community does not exist per se and both a user and a developer community have to be built first. Additionally, firm-initiated SVPs generally do not accept external contributions to the code base unless the contributor transfers the copyright to the vendor (Olson 2005). This is necessary to run dual licensing strategies without fearing legal issues. Since volunteers are often not willing to transfer their copyright, many firm-initiated SVPs exist without a community of active external developers.

In contrast, community-initiated OSSDPs begin with the community, which has no commercial interests. As the project matures and develops commercial potential, an external investor (e.g., software vendor) or the community founders themselves create a company. To generate revenue from dual licensing, a community-initiated SVP also needs to hold the entire copyright for the software product, which might be a threat if not all programmers active in the community migrate to the created company.

As discussed, firms engaging in a MVP do not intend to receive direct revenue from selling the software product. Instead, they contribute to a project along with other parties (typically other firms or voluntary programmers) that also deploy their resources (West & O'Mahony 2008). Thus, the copyright for the developed product is widely distributed. Without ownership of the entire copyright, no party can benefit directly in a commercial sense, such as by pursuing a dual licensing approach. The development approaches, similar to the situation in R&D consortia, might be initiated by the community (e.g., Linux) and include sponsoring firms later or be initiated by a firm that hopes to set a standard and promote followers (e.g., Google's Android).

Previous studies have argued that different modes of collaboration can be assigned to a continuum of integration (Villalonga & McGahan 2005). Thus, I adopt the idea proposed by Van de Vrande, Vanhaverbeke, and Duysters (2009) and rank the approaches in my classification (firm-initiated SVP, community-initiated SVP, firm-initiated MVP, community-initiated MVP) along a continuum from the integration of external parties to market transactions, from the firm's perspective (Figure 4.1).

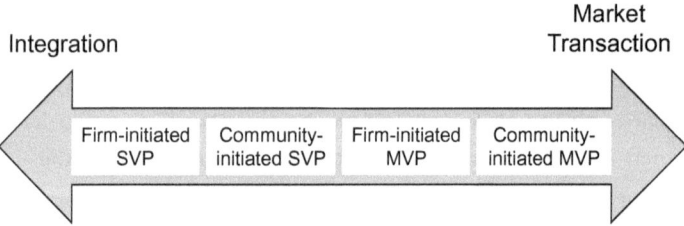

**Figure 4.1** Continuum of OSS Governance Approaches

Firm-initiated SVPs exhibit some similarities to proprietary software vendors (Riehle 2011a). The absence of an active developer community means that SVPs initiated by a firm consist of hierarchical structures. They accept contributions from outside only if they can obtain the copyright for those contributions, which implies a more integrated collaboration mode. Although community-initiated SVPs share the same restrictions with regard to copyright ownership, the community of external developers might be more active and enforce a less integrated mode (O'Mahony 2007).

When MVPs are initiated by a firm, they often seem unattractive to external developers, who do not want to provide their resources for free to serve firm profit goals (Hahn et al. 2008). Unlike SVPs though, other firms participate to ensure the project's continuance, creating more interested stakeholders (De Laat 2007). In MVPs initiated by a community, the firm must manage interactions with other firms *and* integrate a community of external volunteers, who have engaged in the project long before the firm(s) joined. The composition of these different external parties implies a loose coupling of stakeholders, such that community-initiated MVP is the least integrated form of collaboration. These governance modes in turn require different control mechanisms.

# 4.4 Hypotheses Development

The main topics of this chapters are the role of OSS development as a form of R&D alliance and the question of how firms may keep control over OSSDPs in general. According to these two topics, the hypotheses are developed in two steps. In the first step, I formulate hypotheses considering the structure and constitution of OSSDPs with regard to their function as R&D alliances. In the second step, I build upon different forms of OSSDPs and argue that these projects demand different types and intensities of control. Therefore, the hypotheses pertain to the relation of firm-sponsored developers and voluntary developers in relation to different forms of OSSDPs (firm-initiated SVP, community-initiated SVP, firm-initiated MVP, community-initiated MVP) and help to answer questions such as: "Does the number of firms in a project affect the number of sponsored developers (RDBC) and the number of firm-sponsored project leaders (CBL)?"

## 4.4.1 Part I: Open Source and R&D Alliances

As discussed, firm-driven OSSDPs contain natural tensions like the coexistence of collaboration and control. Additionally, from a transaction cost perspective (Williamson 1975), high investments, such as assigning programmers to work for an OSSDP, require a higher level of control than low investments, such as occasionally sponsoring a community. Furthermore, firms prefer non-hierarchical governance when isolating mechanisms such as patents, copyrights, or nondisclosure agreements are effective (Van de Vrande et al. 2006). As products offered under an OSS licence cannot be protected by conventional methods, it is most likely that firms at least want to retain control and to influence development decisions. One way to do so is "having a man on the inside," that is, assigning a firm's own programmers to work in OSSDPs (Dahlander & Wallin 2006).

Simultaneously, as external participants tend to have the wish to (co-) determine development decisions that often compete against firms' interests, firms are facing the challenge to balance the tension between acquiring external knowledge and the wish for control

(Fleming & Waguespack 2007, O'Mahony & Ferraro 2007). As argued, firms that seek to engage in an OSSDP can use different approaches such as sponsoring a server infrastructure, sponsoring committers, or sending their own developers to a specific project. However, it is observable that the majority of firm-driven OSS projects in OSS foundations are run by paid programmers of various firms. As the legality of principle mechanisms explained by transaction cost theory are predicted to hold true for any participating firm, the number of firm-paid programmers is predicted to rise with the number of investing firms. Therefore, it can be hypothesized:

> *Hypothesis 1:* The more firms are involved in a project, the more paid committers a project has.

Literature on governance methods in OSS projects predicts project leadership as one of the most successful control mechanisms (O'Mahony & Ferraro 2007). With regard to the level of intended control, placing firm-paid programmers as project leaders is therefore one of the highest levels of control. However, this logic also holds true for any participating firm. Overall, this would suggest that, in cases with more than one participating firm, namely MVPs, it is most likely that the number of project leaders rises with the number of firms involved.

> *Hypothesis 2:* The more firms are involved in a project, the more paid project leaders a project has.

Looking at the collaborative nature of OSS development, both, firm-paid contributors and voluntary programmers are embedded in a network (e.g., Crowston & Howison 2006, Dafermos 2001, Grewal et al. 2006). In the case of OSS foundations, it is observable that individuals are active in more than one project. Thus, developers are related to each other because they work together on projects and projects are related to one another because they share developers (Grewal et al. 2006). This kind of interdependency is called a two-mode affiliation network (Faust 1997). According to the findings of Dahlander & Wallin (2006), it can be assumed that the role of an individual in a network, typically measured by centrality, differs depending on if he is affiliated with a sponsoring firm or not.[2] If networks of sponsored developers are considered an entity of networks of firms,

---

[2]Social network theory also predicts centrality to be correlated to job performance (Sparrowe, Liden, Wayne & Kraimer 2001, Valverde & Solé 2006). In this study, it was not possible to measure job performance. Although this is a limitation, being sponsored by a firm implies job performance as individuals paid by a firm usually have deeper knowledge on a project that(indirectly) leads to increased job performance.

developers that are sponsored by a firm ought to seek to form ties with other firm-sponsored developers. Furthermore, firm-sponsored developers feature resource advantages since they can devote more time to (even different) projects than their voluntary peers. Thus, based on the above reasoning, being sponsored by a firm affects the position of an individual within a network as follows:

*Hypothesis 3a:* Being sponsored by a firm increases the likelihood that an individual has ties to other individuals.

*Hypothesis 3b:* Being sponsored by a firm increases the likelihood that an individual is close to his neighbors.

Looking at the characteristic of an OSS foundation, it is observable that many firms are active in more than one project. Oftentimes, these projects are linked to each other in the sense that multiple projects contribute to one superior goal. In these cases, a smaller network distance between firms in the network of existing strategic alliances increases the probability that a direct link, when formed, will take the form of a (further) strategic alliance or consortium (Hahn, Moon & Zhang 2008, Hu & Zhao 2009, Vanhaverbeke et al. 2002). Following this logic, emerging firm-initiated projects are often connected to prior projects and share the need for the same kind of knowledge (Grewal et al. 2006). Furthermore, following Méndez-Durón & García (2009), external knowledge sharing among projects through individuals contributing to software development in more than one project will positively influence project success. Therefore, it can be hypothesized, that individuals sponsored by firms are predicted to work on more projects than their voluntary colleagues.

*Hypothesis 4:* Being sponsored by a firm increases the likelihood of participation in more than one project.

### 4.4.2 Part II: Firm Control in Open Source Collaboration

The hypotheses formulated in Part I are comparably general in nature. Therefore, building on different forms of OSSDPs as defined previously (namely firm-initiated SVP, community-initiated SVP, firm-initiated MVP, and community-initiated MVP) in Part II, I will deduce hypotheses regarding different forms of control a firm might most likely execute.

As developed in Section 3.3.4, firms may apply control by leadership (CBL) or resource deployment-based control (RDBC) to influence a project's trajectory. CBL refers to a control mode that is applicable in situations where task programmability and output measurability are given. Instead, RDBC is a concept that refers to clan control outside the boundaries of the firm. In particular, firms assign their own programmers who have been socialized within the company to an OSSDP. As they share norms and beliefs of the company, and moreover, often physically sit in company buildings, it is most likely that they introduce the company's norms and beliefs into the OSSDP.[3]

In addition, a firm's ability to extend its level of control is a function of the resources the firm provides. For example, firms can deploy multiple project leaders in large software projects (e.g., Mockus, Fielding & Herbsleb 2002). If they can capture more leading positions, they also gain more decision-making competence and CBL. In a similar vein, assigning more firm-sponsored developers to a project represents RDBC, which further increases the firm's reputation within the OSSDP (Henkel 2009). For example, Intel and RedHat both provide numerous maintainers, who are allowed to work on the source code, to the Linux kernel.

Previous research concurs that a considerable number of programmers earn pay for their work on OSSDPs (e.g., Lakhani & Wolf 2005, Roberts et al. 2006, Xu et al. 2009). Thus, firms that engage in OSSDPs are likely to exert control to secure their investments. Regarding the role of OSS foundations as boundary organizations, the community might equally use CBL and RDBC in order to influence the project's trajectory and secure its investment (e.g., development time). However, increasing numbers of paid contributors may decrease volunteers' interest in the project and limit the amount of external contributions.

*Hypothesis 5:* Firms make more use of control by leadership and resource deployment-based control than does the community of volunteers in both SVP and MVP.

Although firms commonly prefer to obtain control, those active in an SVP may engage in an OSSDP for reasons different than those that drive firms to be active in MVPs (Fitzgerald 2006, Riehle 2011a, West & O'Mahony 2008). That is, SVPs are similar to a scenario with proprietary software vendors, whereas MVPs are means to collaborate efficiently with other firms, similar to strategic alliances. The presence of multiple firms in an OSSDP and their communal interest in influencing the project implies the number of paid project leaders and committers should be higher for MVPs than SVPs.

---

[3]Various authors have discussed similar situations in the context of "Organizational-Professional-Conflict Theory" (Daniel, Maruping, Cataldo & Herbsleb 2011, Wallace 1995)

*Hypothesis 6:* The number of firm-sponsored project leaders and firm-sponsored committers is higher in MVPs than in SVPs, reflecting an increased need for control by leadership and resource deployment-based control.

Dahlander (2007) further points to the importance of a project's history for different governance modes. The evolution of many OSSDPs has depended on the founder's personality and technical abilities, which influence the composition and activity of the community (Hahn et al. 2008, O'Mahony 2007). Even Linux, one of the big OSS success stories, was long (and still is) dependent on Linus Torvalds, its founder. A project initiated by a community of developers (or a sole influential developer) therefore differs from a project initiated by a firm. To avoid the impression of harnessing a community's effort, firms entering a stable, community-initiated project likely limit their efforts to manage or control the project's trajectory (West & O'Mahony 2008). In addition, the founder's personality might make community-initiated projects more attractive to voluntary programmers (Stewart & Gosain 2006). Therefore, community-initiated projects should consist of more unpaid committers and more unpaid project leaders than their firm-initiated counterparts.

*Hypothesis 7:* The number of voluntary project leaders and voluntary committers is higher for community-initiated projects than for firm-initiated projects, reflecting the community's aim to control firms that enter the OSSDP.

Following Van de Vrande and colleagues (2009), different governance modes can be ranked along a continuum from vertical integration to market transactions (Williamson 1985). Thus, firm-initiated SVPs, community-initiated SVPs, firm-initiated MVPs, and community-initiated MVPs, can also be ranked along a continuum if all forms are treated as governance modes. Firm-initiated SVPs come closest to proprietary software vendors as they internalize external contributions to protect their dual licensing approach. At the other end, community-initiated MVPs are a loose form of collaboration without any legal or intellectual property protections (e.g., Von Hippel & Von Krogh 2003). Thus, the entry barriers for volunteers are highest for firm-initiated SVPs but lowest for community-initiated MVPs, which should reduce the number of voluntary developers in firm-initiated SVPs. The number of firm-sponsored committers might increase though if a company prefers a more integrated governance mode.

*Hypothesis 8a:* The number of voluntary committers decreases with the choice of more integrated governance modes.

*Hypothesis 8b:* The number of firm-sponsored committers increases with the choice of more integrated governance modes.

The multiple firms in MVPs may have diverse views of a project's trajectory (Almirall & Casadesus-Massanell 2010, West & O'Mahony 2008). Considering possible control modes, including the provision of resources to an OSSDP and leadership positions, an increase of firms active in a MVP would imply more firm-sponsored developers and project leaders. However, any firm theoretically can apply each control mode, so the number of firm-sponsored developers and project leaders might increase if more firms commit to the OSSDP. Therefore, it can be finally hypothesized:

*Hypothesis 9a:* In MVPs, the number of firms increases the number of firm-sponsored committers, reflecting resource deployment-based control.

*Hypothesis 9b:* In MVPs, the number of firms increases the number of firm-sponsored project leaders, reflecting control by leadership.

## 4.5 Research Design

### 4.5.1 Research Objective: The Eclipse Foundation

In order to investigate the role of firms in boundary organizations such as OSS foundations, I searched for those foundations that, on the one hand, host a number of projects and, on the other hand, accommodate projects with a high number of participating firms. I found that there are only a few large foundations that fit the requirements, namely Mozilla, Apache, Linux, and Eclipse. I chose Eclipse for several reasons. First, Eclipse started as a project within IBM and is therefore firm-driven and guarantees a high number of participating firms. Second, the foundation is one of the most successful ones with more than 100 members. Third, a number of governance mechanisms are publicly available such as the process of becoming a committer or the responsibilities of its members. A voting system secures the ongoing quality of the committer selection process. See, for example, the following message from the swordfish-news group on news.eclipse.org:

> This automatically generated message signals that Oliver Wolf has nominated Zsolt Beothy-Elo as a Committer on the rt.swordfish project. The reason given is as follows: Zsolt has contributed the initial implementation of the Swordfish service registry and subsequently a number of enhancements. Moreover, Zsolt has agreed to continue

to maintain and further improve the implementation. Zsolt was instrumental in defining the related APIs and the architectural design. The vote is being held via the MyFoundation portal: voters *must* use the portal for the votes to be properly recorded. The voting will continue until either all 5 existing Committers have voted or until they have been given enough time to vote, even if they do not do so (defined as at least one week). Zsolt Beothy-Elo must receive at least three +1s and no -1s for a successful election.

Eligible Committers must cast their votes through their My Foundation portal page (do NOT just reply to this email; your vote will not be correctly recorded unless you use the portal)

Finally, in the case of Eclipse, governance rules ignore the size of a firm. Every strategic board member has only one vote even if they donate much more than the others. Furthermore, the Eclipse website – which is donated by firms such as AMD, HP, IBM, Intel, Magma, and Novell – provides a lot of information concerning Eclipse projects such as the name of every committer along with his affiliation.

The Eclipse foundation, built around the Eclipse project, has a fascinating history. The Eclipse project originated as a development environment within IBM. At that time, its major competitors included Microsoft's Visual Studio and Sun's NetBeans. To gain momentum, IBM open sourced its development, which meant that they were sharing a $40 million investment with competitors (O'Mahony et al. 2005). However, other vendors could build products on top of Eclipse, rather than using proprietary software from competitors. With its open design, the Eclipse project therefore diffused rapidly and attracted much commercial interest. Due to the high number of participating firms along with increasing legal issues, the Eclipse foundation was created to meld different interests and to promote Eclipse.

Today, the Eclipse foundation hosts not only the original Eclipse project but many other projects alike; some of them are just plug-ins, others are elaborated software projects. In general, any contributor is welcome to any project, though potential committers must undergo a process to prove their programming qualifications. Voluntary contributors and participating firms must also agree to certain process rules and a project charter. In addition, every project is based on the principle of meritocracy: The more you contribute and the higher the quality of your contribution, the more you are allowed to do (Young 1994). Finally, project leadership can only be obtained by election.

## 4.5.2   The Process of Data Gathering

The empirical study was based on a data set gathered from the Eclipse website. The advantage of using website data in comparison to email data is that certain limitations, such as the problem of identifying individuals through their email postfix, are obsolete. For instance, every firm affiliation of every committer is unambiguously noted on the website (see Figure 4.2). With the support of students, I developed a database model with tables for

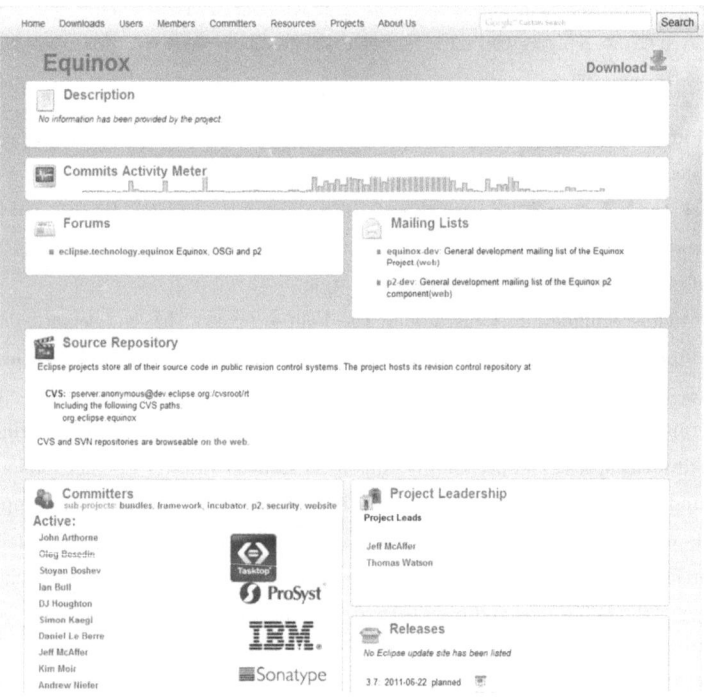

**Figure 4.2** The Eclipse Project Website (Source: www.eclipse.org)

committers, firms, projects, and attributes like the role of every committer in any project, that is, whether he is a project leader or not. The data, though, was primarily collected using a web parser analyzing the Eclipse website in November 2008. Some information had to be added manually to the database via SQL statements like the emergence of two new projects while the parser was working. The advantage of a database is that it is possible to extract different sets of information for different purposes. It is worth noting that the data only mirrors the situation in November 2008 and therefore delivers a static view, not a dynamic one. Finally, the data collection resulted in a set of 912 committers from 110 firms working on 109 projects.

The web presence of the Eclipse foundation is very structured. Participating people are listed by role for different projects and affiliations. I thus could isolate information about each programmer's affiliation and role within a project (committers can be active in multiple projects). Only those who have earned committer status may change parts of the source code. Because the right to change implies an important influence on a project's trajectory, I only consider programmers with committers status and capture the number of voluntary committers, number of voluntary project leaders, and number of firm-sponsored counterparts in each project.

Furthermore, to make use of the measures social network theory is offering, I needed a matrix of who is working with whom. To extract those kinds of data in a matrix, queries were started using SQL statements embedded in php-scripts.[4] In contrast to Grewal et al. (2006), who draw a very precise picture of different persons working on different projects, due to the nature of the method I used, it was only possible to extract if someone was working with someone else or not – ignoring the possibility of multiple collaboration in more than one project. Although this is a limitation in general, it was precise enough regarding the formulated hypotheses in part I (H1-H4).

To test my hypotheses in part II (H5-H9a), I took the same data from the Eclipse website. However, since I was interested in different forms of OSSDPs, I had to develop a coding scheme. In order to distinguish community-initiated from firm-initiated OSSDPs as well as SVPs from MVPs, I first had to identify whether a project was initiated by a firm or a community and then determine the number of firms per project to classify these projects according to my categorization.

I coded projects with more than one participating firm as MVP and those with exactly one participating firm as SVP. I did not find a single project without firm participation, which confirms the perception of Eclipse as a firm-driven foundation. To separate community- from firm-initiated projects, I looked at who submitted the first commit in each project's history. Although each contribution is logged, a recorded commit does not indicate the size of the actual contribution (Arafat & Riehle 2009), so I cannot identify directly if a contributor changed a whole function or just a few lines. Despite this limita-

---

[4]I thank the participants of the research project "Firms in Open Source Software" at University of Koblenz-Landau, namely Martin Braun, Martin Grün, Ibrahim Mat, Patrick Schwirz, and Christopher Felix Wahl for their passionate work and help in data gathering.

tion, many commits should signal firm or community activity and a party's interest in a project.

Of the initial 109 projects, only 83 received a commit in our study timeframe, resulting in a reduced data set for Part II. In a few cases, the first commits came from both developers affiliated with a firm and voluntary developers. I therefore used the number of commits in period $t_0$; for example, if a project was founded in August 2004 and received 500 commits by firms and 20 by volunteers in the first month, I coded it as a firm-initiated project.

### 4.5.3    Operationalization of Variables

Conceptualization and operationalization of variables are important steps in empirical social science. Unlike in cases with latent constructs, the majority of variables needed for my study are directly observable (and measurable), such as the number of paid committers. However, regarding hypotheses H5-H9a, control is a construct that is not directly observable. Consequently, various operationalizations for different forms of control exist. For example, for salespeople, output control is often measured as the ratio between fix and variable income (e.g., Krafft 1999). Instead, behavior control is often measured as the number of touch points a manager has with his subordinates per day (e.g., Ouchi & Maguire 1975).

CBL and RDBC are also latent constructs that require a special operationalization. Thus, the following sections aim at describing where and how observable measures, such as number of firms active in a firm, are interpreted in their purest sense, and where they contribute to unobservable constructs, such as control. In order to separate variables concerning Part I from variables concerning Part II, for Part I variables I use capital letters.

#### 4.5.3.1    Coding of Variables for Part I

**Dependent Variables**    Regarding H1-H4, I concentrated on the following measures. Firstly, it is important to note that due to the nature of H1-H4 there are variables related to a list of projects and variables related to a list of individuals serving H1 and H2, as well as H3a, H3b, and H4, respectively.

Regarding the list of projects, being sponsored by a firm or not is essential for the analysis. Dahlander & Wallin (2006) differentiate between individuals sponsored by a firm with a business model based on the project, individuals sponsored by other organizations, and hobbyists. As projects are not designed with the aim of generating direct revenue, in the case of Eclipse it can be assumed that the difference in behavior of firms with or without an allocated business model is hardly observable. I therefore did not check for the underlying business models of participating firms and coded a committer as paid by a firm if there was a website entry for firm affiliation; otherwise, the person was coded as a volunteer.[5] Thus, the simple numbers of firm-paid (PAIDCOM) and voluntary (VOLCOM) programmers as well as firm-paid project leaders (LEADPAID) on the project level were used as variables.

For H3a, H3b, and H4, the study subject changes from the project to the individual programmer. The dependent variables were derived from social network theory (SNA) (Faust 1997, Wasserman & Faust 1994). In SNA, a network is based on relations (also referred to as ties) between actors. Here, a tie between developer A and B exists if both work together in at least one project. Since ndividuals are central in a network if they have multiple ties to other individuals (Freeman 1979), I therefore coded the number of direct connections to other individuals as degree centrality (DEGCENT). Additionally, I checked if the normalized degree centrality shows any difference (NRMDEGCENT). A second measure for actor centrality is closeness centrality. This measure focuses on how close an actor is to all the other actors (Wasserman & Faust 1994). In contrast to degree centrality, it is the "length" of communication paths to others than the overall number of ties. In this study, this measure is coded as the closeness centrality of an individual programmer (CLOCENT).

Eigenvector centrality is another measure for centrality, showing how close a person is to other well-tied persons. Although this would be of interest for this study since well-tied persons are predicted to be sponsored by firms, I had to drop this measure due to results with a variance near zero. Furthermore, as the dataset consists of symmetric relationships (if person A works with person B, person B works automatically with person A), it is not possible to measure differences between receiving and providing knowledge, which would be possible with Email data (Dahlander & Wallin 2006) and which is proceeded in Chapter 5. Consequently, variables such as in-degree and out-degree are not included here.

**Independent Variables**  For the first study – the observation of the characteristics of a number of projects – the independent variable is simply the total number of participating firms in a specific project (FIRMSTOT). It was generated by counting all firms affiliated

---

[5]Note: By definition, a committer is an active developer who has already shown programming skills and therefore has qualified for committer status.

with the project in the database using the SQL count operator. Regarding the list of individual programmers, for H3-4, it is worthwhile to take a deeper look at firm affiliation. Although the overall number of deviations is very low, I split the group of firm-paid committers in those affiliated with a firm (FIRMPAID) and those working for nonprofit organizations like universities or the Eclipse foundation itself (AFFILIATED). To separate the variable from its interpretation for H1-2 (where it acts as a dependent variable) and to prevent confusion, the group of voluntary committers was renamed to hobbyists for the list of individuals (HOBBYISTS).

**Control Variables**   As mentioned before, from a transaction cost theory perspective, high investments require a high level of control. Therefore, not only the number of firms in a project might have an effect on leadership structures and whether or not a contributor is paid (H1 and H2) but also the size of a project. In this study, size of a project is measured by the total number of individuals working on a project (Méndez-Durón & Garcia 2009), regardless of their affiliation (SIZE).[6]

Furthermore, although the majority of projects is predicted to be lead by firm-paid individuals, some of the projects have a number of individuals without firm affiliation listed in the board of project leaders. This might be the case if a project was not initiated by a firm, but is meanwhile driven by firm activities. I therefore checked for the number of voluntary project leaders in a project (LEADVOL) as well as for the total number of leaders (LEADTOT). Additionally, some OSS cases show, that the maturity of a project affects the structure of a project. More precisely, one stream of literature argues that the older a project is, the more voluntary committers a project attracts due to its publicity (e.g., Bergquist & Ljungberg 2001). On the other side, "project fatigue" could cause unpaid programmers to deallocate resources in later stages of a project (e.g., Walsh et al. 2011). Although studies on factors related to project age or maturity are rare, and, moreover, lead to different results, I include the age of a project as a control variable, measured in month after the official founding date (MATURITY) (cf. Hahn et al. 2008, Sojer 2011).

Literature on social network theory predicts that those people with a more diverse mindset and an extended knowledge base play a special role within a network. In cases of multiple networks so-called boundary spanners are aware of a problem in one net-

---

[6]Although other measures of size would have been possible, such as lines of code, number of individuals working in a project was chosen for practical considerations, that is, availability of data. For hypotheses H5-H9b, size will be coded as number of commits. This also would not have been possible for H1-H2 since commits were not available for the entire set.

work and are therefore able to capture central positions in another network (Dahlander & Wallin 2006, Fleming & Waguespack 2007, Tushman & Scanlan 1981b, Tushman & Scanlan 1981a). A developer's knowledge base, created by working in different projects simultaneously, might impact the decision to elect him as a project leader. In this study, the number of projects an individual is engaged in is an appropriate measure for boundary spanning (NUMPROJECTS). An advanced knowledge base is also reflected by the experience of a programmer. As a limitation of this study, it was not possible to extract information about an individual's experience from the website. As in many other studies on OSS, I excluded gender as a control variable due to the dominance of men in OSS projects (cf. Hertel et al. 2003).

#### 4.5.3.2  Coding of Variables for Part II

Regarding H5-H9b, the majority of figures from Part I will be used. However, as outlined previously, in order to classify different OSS governance approaches such as firm-initiated SVP, community-initiated SVP, firm-initiated MVP, and community-initiated MVP, the number of the firm's contributions (i.e. commits) are a prerequisite. Since commit history was not available for all projects, the dataset was reduced to those projects with a commit history, resulting in 83 projects.

I then interpreted the number of committers as an indicator of RDBC and the number of project leaders as a means to pursue CBL. Similar to prior work on OSSDP, I include project age and size as control variables (Hahn et al. 2008, Hu & Zhao 2009, Stewart & Gosain 2006). In contrast to the coding in part I, for project age, I used the natural log of the month since the first commit; I coded project size as the natural log of the number of commits a project received, to compensate for skewness (Hahn et al. 2008).

All other figures, which are treated as variables in Part I, remain unattached except for the fact that the dataset was reduced.

# 4.6   Results

## 4.6.1   Part I: Open Source and R&D Alliances

As mentioned in the introduction, the process of data gathering revealed 912 committers, 109 projects, and 110 firms. Each project receives contributions by at least one firm and a majority of projects are led by at least one paid project leader (Figure 4.3). I used

SPSS version 18 for all statistical calculations such as correlations and regression analysis (Brosius 2006, Field 2005). Table 4.2 shows the descriptive statistics and correlations for the characteristics of projects. As expected, the number of paid committers per project (mean of 11.98) is much higher than the number of voluntary committers (mean of 2.14). This indicates that the projects are definitely characterized by high firm involvement. Accordingly, the number of paid project leaders (mean of 1.37) is higher than the number of voluntary, non-paid project leaders per project (mean of 0.18). However, the results also indicate that there might be a problem with multicollinearity due to the nature of the coding scheme. In cases with many firm-paid committers on a project (in ratio to unpaid committers), their number highly correlates with the overall number of committers, which was coded as project size.

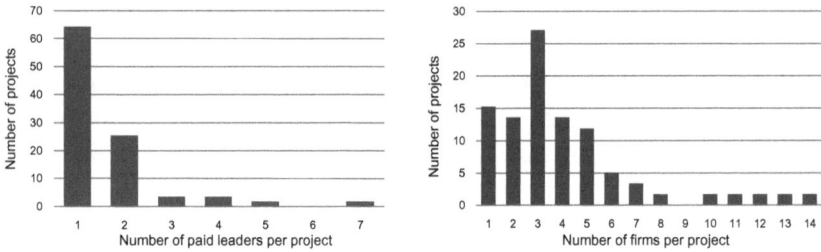

**Figure 4.3** Distribution of Paid Project Leaders and Firms Per Project

Table 4.3 shows the descriptive statistics and correlations for different individuals working on projects. As in Table 4.2, the dominance of firm participation is obvious (mean of 0.77 for FIRMPAID). Figure 4.4 shows the visualization of the network of developers using Pajek (Borgatti, Everett & Freeman 2002). The structure looks very monolithic and less modular than visualizations to be found in other studies (Dahlander & Wallin 2006, Grewal et al. 2006). This also supports the initial assumption that OSS development has moved from the bazaar style to more structured approaches, close to pure firm projects. Furthermore, the network's density is relatively low (d=0.0378, not reported in a table), which points to a loose coupling between individuals.

In Model 1A and 1B, the number of paid committers is the dependent variable (Table 4.4). The difference between Model 1A and 1B is the consideration of project size as a control variable. Model 1B shows a significant and positive effect of the number of firms on the number of firm-paid committers, which supports H1. However, the full model including

**Table 4.2** Descriptive Statistics and Correlations (Based on Projects)

| Variables | N | Mean | S.D. | 1 | 2 | 3 | 4 | 5 | 6 | 7 |
|---|---|---|---|---|---|---|---|---|---|---|
| 1 MATURITY | 97 | 35.35 | 21.06 | | | | | | | |
| 2 PAIDCOM | 109 | 11.98 | 14.73 | .034 | | | | | | |
| 3 VOLCOM | 109 | 2.14 | 3.19 | −.016 | .213* | | | | | |
| 4 SIZE | 109 | 14.12 | 15.73 | .028 | .979** | .407** | | | | |
| 5 FIRMSTOT | 108 | 3.19 | 2.11 | .077 | .590** | .171 | .587** | | | |
| 6 LEADPAID | 109 | 1.37 | 1.21 | .084 | .203* | −.216* | .145 | .202* | | |
| 7 LEADVOL | 109 | .18 | .47 | .000 | −.204* | .478 | −.092 | −.127 | −.070 | |
| 8 LEADTOT | 109 | 1.55 | 1.27 | .082 | .117 | −.028 | .104 | .145 | .928** | .306** |

Note: $^*p < .05$; $^{**}p < .01$; $^{***}p < .001$

**Table 4.3** Descriptive Statistics and Correlations (Based on Individuals)

| Variables | N | Mean | S.D. | Min | Max | 1 | 2 | 3 | 4 | 5 | 6 |
|---|---|---|---|---|---|---|---|---|---|---|---|
| 1 NUMPROJECTS | 872 | 1.77 | 1.52 | 1 | 16 | | | | | | |
| 2 DEGCENT | 910 | 34.54 | 38.07 | 0 | 192 | .534** | | | | | |
| 3 NRMDEGCENT | 910 | 3.78 | 4.18 | 0 | 21.07 | .534** | 1.000*** | | | | |
| 4 CLOCENT | 900 | 0.43 | 0.16 | 0.11 | 0.51 | .199** | .265** | .266** | | | |
| 5 FIRMPAID | 910 | 0.77 | 0.42 | 0 | 1 | .126** | .127** | .126** | -.080* | | |
| 6 AFFILIATED | 910 | 0.03 | 0.18 | 0 | 1 | -.052 | -.085** | -.085** | .050 | -.344** | |
| 7 HOBBYISTS | 910 | 0.20 | 0.39 | 0 | 1 | -.109** | -.096** | -.095** | .062 | -.903** | -.093** |

Note: $*p < .05$; $**p < .01$; $***p < .001$

the control variable provides a different picture. To check for multicollinearity between the independent variable and control variable, I calculated the variance inflation factor (VIF) (Hair, Anderson, Tatham & Black 1998). Although VIF scores for multicollinearit between independent variables achieved results below the recommended threshold of 10, collinearity between independent and dependent variable seems to be an issue in Model 1A, which is also mirrored by a very high $R^2$ value.

**Figure 4.4** Interactions of Individuals Working on Eclipse Projects

Serving H2, in Model 2A, the number of voluntary committers has a negative and significant effect on the number of firm-paid leaders whereas size has a positive effect. Obviously, these effects explain more dependencies than the number of firms involved. Removing these measures leads to Model 2B where the number of firms has a slight positive effect on the number of leaders. Overall, the figures show only weak support for H2.

Table 4.5 shows the effect of sponsoring on the individual level. The number of projects an individual is working on (NUMPROJECTS) was intended to be a control measure as participation in more than one project points to advanced experience. Again, according to other studies (e.g., O'Mahony & Ferraro 2007), working on many projects might be an indicator for boundary spanning.

**Table 4.4** The Effect of Firm Engagement on Projects

|  | Model 1A DV=PAIDCOM | Model 1B DV=PAIDCOM | Model 2A DV=LEADPAID | Model 2B DV=LEADPAID |
|---|---|---|---|---|
| VOLCOM |  |  | -.350***(.001) |  |
| MATURITY | .005(.835) | -.008(.929) | .063(.507) | .067(.504) |
| SIZE | .961***(.000) |  | .341**(.006) |  |
| FIRMSTOT | .026(.339) | .539***(.000) | .083(.464) | .220*(.031) |
| Number of projects | 97 | 97 | 97 | 97 |
| R-squared | .951 | .291 | .183 | .151 |

Note: *$p < .05$; **$p < .01$; ***$p < .001$; Standard errors in parentheses

**Table 4.5** The Effect of Sponsoring on Individuals

|  | Model 3A DV=DEGCENT | Model 3B DV=DEGCENT | Model 4A DV=CLOCENT | Model 4B DV=CLOCENT |
|---|---|---|---|---|
| NUMPROJECTS | .534***(.000) | .525***(.000) | .199*(.000) | .213***(.000) |
| FIRMPAID |  | .055**(.006) |  | -.093**(.009) |
| AFFILIATED |  | -.037(.228) |  | .030(.394) |
| Number of individuals | 872 | 872 | 872 | 872 |
| R-squared | .285 | .290 | .040 | .051 |

Note: *$p < .05$; **$p < .01$; ***$p < .001$; Standard errors in parentheses

However, as firm-paid contributors work, on average, on more projects than their voluntary colleagues, the effect of being on more than one project on network position (i.e., DEG-CENT and CLOCENT) is very strong and significant. More precisely, Model 3A shows the single dependency between the number of projects a person is involved in and degree centrality. The effect of sponsoring is positive and significant (as can be seen in Model 3B). However, it is not very strong and the effect of being active in many projects seems to preponderate. Despite the moderating effect of number of projects a person is involved in, the findings provide support for H3a. Similar influences are observable in Model 4A and 4B. As in the case of degree centrality, the moderating effect of NUMPROJECTS is strong for closeness centrality. Surprisingly, the effect of sponsoring, namely the influence of being sponsored by a firm, is slightly negative. However, the coefficient of determination $R^2$ for Model 4A and 4B provides values of .04 and .051, respectively, which indicates that

only less than 5% of the variance is explained by the model, a result of the low variance
for CLOCENT (Table 4.3). H3b therefore has to be rejected.

In order to investigate the differences between the types of sponsoring, I report the
mean values of the dependent variables in Table 4.6 (cf. Toutenburg & Heumann 2008).
As expected, in support of H4, firm-paid contributors are more central and participate
in more projects than their voluntary (HOBBYISTS) and non-sponsored (AFFILIATED)
colleagues.

**Table 4.6** Comparing Means of Dependent Variables

| Variable | FIRMPAID | AFFILIATED | HOBBYISTS | F-Test |
|---|---|---|---|---|
| NUMPROJECTS | 1.87 | 1.35 | 1.42 | 7.008(.001) |
| DEGCENT | 37.1444 | 17.1935 | 27.1012 | 8.378(.000) |
| CLOCENT | .4187 | .4675 | .4457 | 3.159(.043) |
| Number of individuals | 701 | 31 | 178 | 912 |

### 4.6.2  Part II: Firm Control in Open Source Collaboration

H5-H9b require the judgment of whether or not a project was initiated by a firm (Dahlander
2007). As discussed, the only possibility to check for initiation was to identify which party
made the first contribution (i.e., commit). After checking for projects with little or no
activity and ones without any commits, we ended up with a list of 83 projects. Of these
projects, 17 were single vendor projects, 66 were multi vendor projects, 33 were community
initiated, and 50 were firm initiated. Figure 4.5 illustrates different commit behaviors in
single and multi vendor projects with the timeline on the x-axis and number of commits
on the y-axis.[7]

As a first step, I conducted an analysis of variance (ANOVA) to compare the mean
values for each of the categorical variables (i.e., initiation and vendor type) (Cohen 1988).
Specifically, I examined whether the mean values for the number of commits of firms, of
volunteers, and of paid project leaders in a project differ between different approaches. To
run the ANOVA, I split the sample of projects into single vendor and multi vendor projects
for a first analysis (Table 4.7) and into firm-initiated and community-initiated for a second
analysis (Table 4.8).

---

[7]The absolute numbers as well as the names of participating firms are not of interest here. The figure
just portrays that in MVPs more firms commit than in SVPs and therefore provides support for the
distinction between both types of OSSDPs.

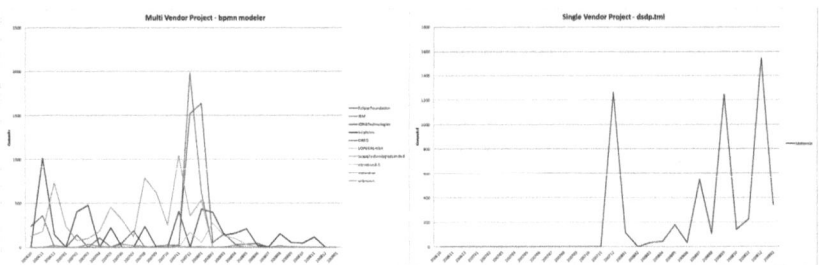

**Figure 4.5** A Comparison Between a Selected Multi (left) and a Single Vendor Project (right)

The results show that MVPs receive significantly more commits than SVPs. Furthermore, it is worth noting that the number of paid committers in MVPs is nearly three times the number of committers in SVPs. As expected, the distribution of voluntary and paid project leaders also significantly differs between SVPs and MVPs. Moreover, the overall number of voluntary project leaders is indeed very low (mean of 0.35 in SVP, 0.12 in MVP), confirming the commercial culture of Eclipse projects. To further illustrate differences between SVP and MVP, I calculated the ration of paid to all committers. Here I find MVPs to feature 87.54% paid committers in contrast to 76.06% in SVPs.

Regarding Table 4.8, community-initiated projects receive significantly more commits by volunteers than their firm-initiated counterparts. Interestingly, the number of commits by firms does not differ between community- and firm-initiated projects. It is also observable, that the number of paid committers in firm-initiated projects is nearly twice the number of paid committers in community initiated ones. Considering the leadership structure in both types of OSSDP, I expected the number of paid project leaders to be significantly higher in firm-initiated projects. However, the comparatively high number of paid project leaders in community initiated projects was especially a surprise.

With respect to H5-H9b, I employed a multimethod data analysis approach; a single method could not capture all dimensions of the hypotheses. To test H5-H7, I first conducted a multivariate analysis of variance (MANOVA) to compare the mean values for each of the four categories for RDBC and CBL as an extension of the ANOVA presented above.

**Table 4.7** Means (M) and Standard Deviations (SD) After ANOVA on Factor VEN-DORTYPE

|  | Single-vendor project | | Multi-vendor project | |
| --- | --- | --- | --- | --- |
|  | M | SD | M | SD |
| Number of commits by firms | $7.41^a$ | n.c. | $9.06^b$ | n.c. |
| Number of commits by volunteers | $4.08^a$ | n.c. | $5.05^a$ | n.c. |
| Number of all commits | $8.63^a$ | n.c. | $9.37^a$ | n.c. |
| Number of voluntary committers | $2.12^a$ | 3.57 | $1.80^a$ | 2.63 |
| Number of paid committers | $5.18^a$ | 2.90 | $14.83^b$ | 15.92 |
| Number of voluntary project leaders | $.35^a$ | .61 | $.12^c$ | 0.45 |
| Number of paid project leaders | $.88^a$ | .60 | $1.58^b$ | 1.16 |
| Ratio paid to all committers | $76.06^a$ | 35.13 | $87.54^b$ | 17.14 |

**Notes:** Mean values reported, $N = 82$. Within each row, means with a *different* superscript are not significantly ($p < 0.05$ for b; $p < 0.1$ for c) different from each other. n.c.=not calculated by SPSS

**Table 4.8** Means (M) and Standard Deviations (SD) After ANOVA on Factor INITIATION

|  | Community initiated | | Firm initiated | |
| --- | --- | --- | --- | --- |
|  | M | SD | M | SD |
| Number of commits by firms | $8.10^a$ | n.c. | $8.95^a$ | n.c. |
| Number of commits by volunteers | $6.07^a$ | n.c. | $4.03^b$ | n.c. |
| Number of all commits | $9.26^a$ | n.c. | $9.11^a$ | n.c. |
| Number of voluntary committers | $2.36^a$ | 3.53 | $1.60^a$ | 2.26 |
| Number of paid committers | $8.45^a$ | 6.44 | $15.46^b$ | 17.77 |
| Number of voluntary project leaders | $.21^a$ | .48 | $.16^a$ | .51 |
| Number of paid project leaders | $1.33^a$ | 1.13 | $1.48^a$ | 1.09 |
| Ratio paid to all committers | $79.36^a$ | 29.50 | $86.94^a$ | 19.59 |

**Notes:** Mean values reported, $N = 83$. Within each row, means with the *same* superscript are not significantly ($p < 0.05$) different from each other. n.c.=not calculated by SPSS

I ran a MANOVA instead of several independent ANOVAs to reduce the type-I error that emerges from multiple tests with the same data set (Field 2005, Podsakoff, MacKenzie, Lee & Podsakoff 2003, Stevens 2002) and to be able to investigate interaction terms.

A 2×2 MANOVA with RDBC (number of voluntary committers versus number of firm-sponsored committers) and CBL (number of voluntary project leaders versus number of firm-sponsored project leaders) as dependent variables and initiation (firm-initiated versus community-initiated) and vendor type (SVP versus MVP) as independent variables reveals significant main effects (initiation Wilks' Lamda = .866, $F_{(6,73)}$ = 1.88, $p < .1$, $\eta = .134$; vendor type Wilks' Lamda = .845, $F_{(6,73)}$ = 2.22, $p < .05$, $\eta = .155$). However, there is no significant effect of the multivariate interaction (initiation vendor type Wilks' Lamda = .899, $F_{(6,73)}$ = 1.36, sig.= .245, $\eta = .101$), as I show in Table 4.9.

**Table 4.9** Means (M) and Standard Deviations (SD) After MANOVA

| | | Vendor type | | | |
|---|---|---|---|---|---|
| | | SVP | | MVP | |
| Variable | Initiated by | M | SD | M | SD |
| Number of voluntary committers | Firm | $2.33^a$ | 3.57 | $1.44^a$ | 1.88 |
| | Community | $1.87^b$ | 3.83 | $2.42^c$ | 3.54 |
| Number of firm-sponsored committers | Firm | $5.00^a$ | 2.24 | $17.76^b$ | 18.86 |
| | Community | $5.38^a$ | 3.66 | $9.83^c$ | 6.72 |
| Number of voluntary project leaders | Firm | $.33^a$ | .50 | $.12^b$ | .51 |
| | Community | $.38^a$ | .74 | $.13^b$ | .34 |
| Number of firm-sponsored project leaders | Firm | $.88^a$ | .64 | $1.54^b$ | 1.21 |
| | Community | $.89^a$ | .60 | $1.61^b$ | 1.14 |

**Notes:** For each consumer group, means sharing a common superscript notation within each row and column do not differ; means with different superscripts are different from one another ($p < .05$).

Consistent with H5, firms in general (both in MVPs and SVPs) have significantly more project leaders and committers than the surrounding community of volunteers. However, the multivariate interaction between initiation and vendor type is not significant, so I consider H5 only partially supported. Concerning H6, MVPs consist of more firm-sponsored project leaders and committers than do SVPs. In line with H7, community activity is higher in community-initiated OSSDPs.

As the basis for H8a and H8b, I considered the possibility of ranking different governance modes along a continuum from firm-initiated SVPs to community-initiated MVPs. I initially presumed that an ordered logistic regression would be an appropriate method for controlling for the influence of an ordered dependent variable; however, a Wald test (Brant 1990) indicates the violation of some critical prerequisites. Similar to other investigations of different governance modes as dependent variables (e.g., Van de Vrande et al. 2009), I find that the predicted ranking is more complicated than might be anticipated. More factors refer to the choice of governance modes than those I consider in this investigation. Therefore, I tested the hypotheses with a multinomial logit model. Although in principle, many of these results could be achieved with an ANOVA as well, multinomial logistic regression can include control variables (Hosmer & Lemeshow 2000). I therefore estimated two models, one with independent variables only (Model 1) and one with the dependent variables included (Model 2). I provide the results in Table 4.10.

To present the results, I use community-initiated MVPs as the default category. In H8a, I argued that the number of voluntary committers decreases with the choice of more integrated governance modes. In a similar vein, I posit in H8b that the number of sponsored committers increases with the choice of more integrated governance modes. From Table 4.10, I find partial support for my hypotheses. That is, in Model 1, I estimate the choice of governance modes without including the variables of interest, sponsored committers, and sponsored project leaders, and do not observe any significant effect.

In Model 2, as I predicted, firm-initiated MVPs differ significantly from the default category of community-initiated MVPs in terms of the number of voluntary (fewer) and firm-sponsored (more) committers. The other independent and the control variables show no significant effects. Furthermore, for Model 2, the values of the Nagelkerke and Cox/Snell R-square indicate a good model fit, in partial support of H8a and H8b (Field 2005). The effect of a change in the number of committers – and therefore in the use of RDBC – is not observable for SVP governance modes though. Perhaps the significantly fewer number of cases of SVP, compared with MVP, in my data set produces this result.

Finally, I used ordinary least square (OLS) regression (Aiken & West 1991) to test H9a and H9b (see Table 4.11). A metric scale for all variables is a precondition for applying OLS regression, which is featured by the investigated data set. My estimation shows that the number of firms in a project does not affect the number of firm-sponsored project leaders.

Table 4.10 Multinomial Logistic Regression

|  | Model 1 | | | Model 2 | | |
|---|---|---|---|---|---|---|
|  | SVP firm | SVP com | MVP firm | SVP firm | SVP com | MVP firm |
| Constant | 5.444** | −3.176 | 1.274 | 5.993** | −4.058 | 2.601 |
| Number of voluntary committers | −.215 | −.190 | −.174 | −.279 | −.466 | −.394** |
| Number of firm-sponsored committers |  |  |  | −.171 | −.198 | .121** |
| Number of voluntary project leaders | .919 | 1.407 | .550 | .846 | 1.585 | 1.487 |
| Number of firm-sponsored project leaders |  |  |  | −.456 | −1.389 | −.208 |
| Project age | −1.180* | .094 | −.448 | −1.213* | −.315 | −.675 |
| Project size | −.270 | .193 | .112 | −.128 | .798 | −.046 |
| $\chi^2$ | 16.741 |  |  |  | 47.390 |  |
| -2 Log likelihood | 176.08 |  |  |  | 145.43 |  |
| Cox/Snell Pseudo $R^2$ | .185 |  |  |  | .439 |  |
| Nagelkerke Pseudo $R^2$ | .204 |  |  |  | .485 |  |

Notes: The community-initiated MVP is the comparison group.
$N = 83$; ***$p < .01$; **$p < .05$; *$p < .1$

**Table 4.11** OLS Regression

|  | Model 1<br>DV = number of firm-sponsored<br>project leaders | Model 2<br>DV = number of firm-sponsored<br>committers |
|---|---|---|
| Number of firms involved | .108 | .471*** |
| Project age | .113 | .044 |
| Project size | .111 | .094 |
| $R^2$ | .049 | .243 |
| $R^2$ adjusted | .002 | .206 |

$N = 83$; ***$p < .01$; **$p < .05$; *$p < .1$

**Table 4.12** Summary of Hypotheses

| Hypothesis | Relation tested | supported |
|---|---|---|
| H1 | FIRMSTOT $\xrightarrow{+}$ PAIDCOM | yes |
| H2 | FIRMSTOT $\xrightarrow{+}$ LEADPAID | partially |
| H3a | FIRMPAID $\xrightarrow{+}$ DEGCENT | yes |
| H3b | FIRMPAID $\xrightarrow{+}$ CLOCENT | no |
| H4 | FIRMPAID $\xrightarrow{+}$ NUMPROJECTS | yes |
| H5 | CBL and RDBC is higher for firms in both SVP and MVP | partially |
| H6 | CBL and RDBC (Firm) is higher for MVP | yes |
| H7 | CBL and RDBC (Community) is higher for community-initiated OSSDP | partially |
| H8a | Number of voluntary committers $\xrightarrow{-}$ firm-initiated SVP | partially |
| H8b | Number of paid committers $\xrightarrow{+}$ firm-initiated SVP | partially |
| H9b | Number of firms $\xrightarrow{+}$ RDBC, MVP only | yes |
| H9b | Number of firms $\xrightarrow{+}$ CBL, MVP only | no |

However, in Table 4.11, Model 2, which uses the number of firm-sponsored committers as the dependent variable, shows a significant effect of the number of firms on the use of RDBC. The beta coefficient for the number of firms and number of firm-sponsored committers is .471 and highly significant; when the number of firms changes by one standard deviation, the estimated outcome variable of firm-sponsored committers changes by .471 standard deviations, on average. Model 2 further indicates the adjusted R-square equals .206, in robust support of H9a. I must reject H9b, regarding the use of CBL. Table 4.12 provides a summary of all hypotheses for parts I and II.

## 4.7     Discussion and Conclusion

As argued in the introduction, today, OSS is no longer a programming ideology driven by altruistic hobbyists – if it ever was. Moreover, with the participation of firms it has become a business. For firms, being active in OSS projects has several advantages such as the acquisition of external innovative ideas at more or less no cost. However, to appropriate efficiently, firms have to balance the tension between control and collaboration and this balance may feature different forms in different governance modes.

In light of firm presence in OSSDPs, the overall setting of this research was to show that firms not only are active in OSSDPs but are able to form consortia-equivalent structures under an OSS umbrella. The main advantage compared to well-known collaboration approaches like strategic alliances or classical R&D consortia lies in the openness and flexibility of the OSS approach. Whereas classical approaches have to deal with questions of IP and other legal aspects (Lichtenthaler 2009b, Lichtenthaler & Ernst 2007), using the OSS model leads to products available for everyone and invites new participants to join a project. In this environment, OSS foundations deliver the required framework for firms to establish R&D consortia, especially in cases of low competition between participants.

The empirical findings further endorse the evolution of OSS from a development approach to an inter-firm collaboration alternative. In the case of Eclipse projects compared to other OSSDPs, on average, the total number of individuals paid by firms is relatively high. For instance, Dahlander & Wallin (2006) report a mean for individuals sponsored by firms with a business model dedicated to the OSS product with 0.023 for the GNOME project.

However, the mean of FIRMPAID in this study (which is based on a similar coding scheme) is reported with 0.77. Thus, the investigation further shows that, compared to other projects such as GNOME, Eclipse projects are mainly driven by firms and that only a few contributions from external (voluntary) individuals occur.

Speculating upon those findings, it is most likely that in professional settings firms seek participation of other firms rather than contributions from hobbyists. If a firm decides to establish a business model around a certain OSS project or intends to share resources as in an R&D consortium, relying on altruistic programmers who occasionally commit some lines of code is impossible. Consequently, the majority of the successful OSS projects, such as the Linux kernel, are run by hierarchical governance mechanisms. The pure OSS approach has its limits when it comes to commercialization. Despite all advantages mentioned before, starting an OSSDP as a firm with the hope of getting hundreds of developers at almost no cost will definitely remain a myth (Goldman & Gabriel 2005). Firms are not searching for hundreds of voluntary committers. Instead, they are searching for the few good ones. Not surprisingly, those can be found within competing firms.

Whereas former studies treat the appearance of firms in OSSDPs as an impurity and focus on the role of the community (West & Lakhani 2008), this study shows that OSS development as in the case of Eclipse is far from being community-driven. Especially if hosted under the umbrella of a foundation, the development style of a project is coupled with strict rules and therefore very structured. In the case of Eclipse, the project organization is therefore far from being a bazaar style for which OSS approaches were long famous (Demil & Lecocq 2006, Raymond 1998).[8] Thus, the presence of firms requires a dyadic approach to invite external participation while at the same time influence development decisions.

As argued, Eclipse projects possess many characteristics of R&D consortia. However, what is surprising is that there are voluntary programmers and even voluntary project leaders at all. It can be assumed that those projects have a different history. It is most likely that they were started by unpaid individuals and taken over by firms. Oftentimes, the founder of a project has enormous influence over a group of developers. Obviously, firms want to avoid destroying these grown governance mechanisms within communities. Certainly, this question can only be answered by further research.

Regarding Part II of my analysis, understanding how to influence and control the development of an OSSDP is vitally important to firms that provide resources to that project. Drawing on a behavioral view of control, I argue that firms may choose between RDBC, which authorizes developers socialized within firm boundaries to work for an OSSDP, and

---

[8] Berdou (2011) speaks of a gift economy inside an exchange economy.

CBL, in which case firm-sponsored developers capture leading positions for a project. In addition, I distinguish OSSDPs initiated by a firm or a community, as well as those that consist of one participating firm (SVP) versus multiple firms (MVP). The difference between SVP and MVP is important and often neglected in studies of OSSDPs, yet these approaches reflect different business models (e.g., Dahlander 2007, Riehle 2011b, West & O'Mahony 2008).

In my sample, the difference between SVP and MVP, in terms of the number of firm-sponsored and voluntary developers and project leaders, is more explicit than the distinction based on who initiated the project. Because SVPs mirror proprietary software vendors and their use of a dual licensing approach (Watson et al. 2008) whereas MVPs are a means to collaborate efficiently in a consortium, the choice of a business model clearly determines the choice of control modes.

In my investigation, I draw on transaction cost economics and argue that different governance modes (firm-initiated SVP, community-initiated SVP, firm-initiated MVP, and community-initiated MVP) represent a continuum that reflects the number of transactions. Among MVPs, community-initiated ones differ from firm-initiated ones. As predicted, the number of firm-sponsored committers is significantly higher in firm-initiated MVPs, but the number of voluntary committers is significantly lower. I find no such effect for the group of SVPs. I had a limited number of SVPs in my sample, and the size of a project also might influence these results.

Probably the most interesting finding from my study is that in MVPs, the use of RDBC is more prevalent than the use of CBL. When any new firm enters an OSSDP, the number of stakeholders grows. In theory, the de novo entrant deploys equal resources to the OSSDP as that firm also intends to influence the project's trajectory. I have argued that in those cases, the number of firm-sponsored committers and project leaders increases the ability to perform both RDBC and CBL. However, my results indicate only an increase in RDBC; the number of firms active in an OSSDP does not appear to affect the number of project leaders.

With the data, I cannot determine which form of control is superior in terms of outcomes either – I can just say that RDBC seems to be more common. Therefore, I am cautious about implying that RDBC is a "superior" form of control in MVPs. Instead, I find two main effects that lead to greater RDBC. First, when firms enter a project late, it is very difficult for them to capture leadership positions, because leadership often derives from advance contributions to a project (Dahlander & O'Mahony 2011, Giuri et al. 2008, Scozzi et al. 2008). Therefore, de novo entrants either must wait until their own developers reach leadership positions or hire current project leaders. The latter tactic would not increase the number of project leaders. Second, more project leaders may simply lead to greater coordination costs. If increased coordination costs outweigh the benefits of the division of labor (Aiken & Hage 1968), CBL would not be applicable, even if it seems appealing from a single firm's perspective. This also explains why H2 is not supported. Rather, they have no choice other than to assign their developers to the OSSDP to exercise control, which I define as RDBC.

As does almost any research, my study involves several limitations. First, regarding the data set, Eclipse is a firm-driven environment. Studies of projects from a random sample might reveal different results. Second, my data set does not allow me to separate the different firms in MVPs, so I only considered firm-sponsored individuals, regardless of which firm sponsored them. Further research should instead take into account the influence of different firms in OSSDPs. Third, from a theoretical point of view, control includes both setting directions ex ante and ex post monitoring in a recursive way. Employees receive advice, to which they respond by completing their task, which a supervisor then evaluates. I could not differentiate between ex ante and ex post control mechanisms, so I call for research that considers the time that each control mode is used.

In light of my results, other important theoretical considerations might add to the understanding of the role of firms in OSSDPs. For example, the division of labor is often considered a benefit of using OSSDPs as complementary assets. Dahlander & Magnusson (2008) show that creative work tends to be performed in the community, whereas routine tasks take place within firm boundaries. This study therefore offers new avenues for research that investigates different governance approaches on task level.

Finally, my study has several implications for management. Firms that plan to engage in existing OSSDPs or that are willing to initiate their own projects should define their business model before determining which governance modes to use. According to my investigation, de novo entrants in existing projects prefer RDBC over CBL. Thus, instead of expending resources to hunt for them or granting single developers the time to climb the meritocracy ladder within the community (Henkel 2009), firms should assign many of their own developers to the OSSDP to encourage the transfer of their own norms into the project. These developers may act like a clan within the OSSDP, moving their (and the firm's) preferred topics to the top of the agenda.

# Chapter 5

# Open Source in Action II: Business Collaboration Within an Open Source Project

## 5.1 Introduction

As discussed throughout the dissertation, firms are able to profit from technological developments without having ownership. However, as with any public good, the absence of ownership of the invention means the innovating firm has limited options of protecting what they do not own against unintended use by legal means. Admittedly, they might obtain control over the project by assigning their own developers to work for an OSSDP. In turn, developers working in an OSSDP who earn salaries from the firm are also simultaneously embedded in organizational settings of the firm for which they work (Henkel 2009). By inserting their norms and beliefs, which partly reflect the employing firm's interest, into the OSSDP, these employees allow firms to (indirectly) influence the project's trajectory – depending on the number of programmers assigned and their role on the project. Dahlander & O'Mahony (2011) consequently note that firms do not necessarily need to gain control over a project as long as they control individuals working on that project. This was the basis for the introduction of CBL and RDBC as means to supervise a project's development.

In the case of community-initiated MVPs, influencing the development of the OSSDP is even more challenging. First, as other forms of virtual and non-virtual communities, such as virtual organizations (Davenport & Daellenbach 2011) or consulting companies

(Alvesson & Robertson 2006), OSS community members share certain norms and beliefs. By accepting and promoting a special culture of free access, free code sharing, and voluntary participation, they identify with their community-initiated OSSDP (Albert, Ashforth & Dutton 2000, Von Hippel & Von Krogh 2003). This culture is said to be a condition for trust as a governance mechanism. With the emergence of firm participation, this culture will possibly be violated. Whereas programmers who were active in a project from the beginning still share the original culture, late entrants may mirror firm interests that seem to be contradictory to the OSS ideology.[1] Therefore, firm participation in OSS communities results in a number of tensions like the one between collaboration and control or divergence and discovery (Almirall & Casadesus-Masanell 2010).

Second, recent research has further highlighted that in the case of OSS, the community of developers may be considered as a complementary asset (e.g., Dahlander & Wallin 2006). Following Teece (1986), if the complementary asset is critical to the firm, internalization is the most suitable strategy. In this sense, assigning a firm's own developers to work for an OSSDP is a – rather abstract – form of internalizing externalities. However, the degree of necessary internalization is dependent on the business model the firm applies. Accordingly, Pisano & Teece (2007, p. 287) emphasize that:

> a weakening of the appropriability regime through the emergence of open source operating systems can be beneficial to companies (like IBM) with strong downstream asset positions in middleware, applications, hardware, and services, while damaging firms (like Microsoft) with strong positions in operating systems.

Consequently, firms are likely to engage in the development of OSS to different degrees. Whereas firms with a business model dependent on the OSSDP will assign a larger proportion of developers to the project and seek to obtain leadership positions, firms that only complement their business with OSS will do so to a lesser degree. Thus, firms that intend to influence an community-initiated MVP face a two-part management challenge in that they not only have to align with a community of volunteers and other firms but with firms pursuing different degrees of OSS engagement.

Various authors have noted that by interacting with each other, OSS developers constitute what commonly is considered a network (e.g., Fershtman & Gandal 2011, Iannacci &

---

[1]Indeed, Berdou (2007, p. 88) reported on the basis of interviews with KDE developers that "proprietary developers that are brought in to work on community projects have to learn the ways of the community and adjust to the rhythms and the demands of F/OS development." Moreover, the majority of interviewees thought the group of commissioned late entrants will have difficulties in aligning with the community's culture and way of working.

Mitleton-Kelly 2005, Singh & Tan 2010, Tuomi 2001, Xu, Christley & Madey 2006). Moreover, since firms are represented by their fellow employees, developers within an OSSDP also build what might be considered an interfirm network that enables firms to exchange knowledge through their employees (Knox, Savage & Harvey 2006, Uzzi & Lancaster 2003). Thereby an organization can be represented by a single individual or by multiple individuals.[2] As in any network, central network positions are affiliated with power and (informal) leadership. Actors that are central in a network interact with more individuals than their peers and therefore put themselves in a position to coordinate tasks. Since coordination work in a project is valued by peers, the coordinating actor attracts supporters over which he has authority (Berkowitz 1956, Bonacich 1987, Shaw 1964).

In line with this literature, recent investigations of OSSDPs revealed that an individual's position within a network of distributed programmers is important for the perception of his leadership abilities (e.g., Dahlander & O'Mahony 2011, Fleming & Waguespack 2007, Giuri et al. 2008). Since firms seek to receive CBL by capturing leadership positions within the project, firm-sponsored developers might therefore obtain central positions within the network. Moreover, individuals sponsored by a firm with a business model dedicated to OSS ought to be even more central than other sponsored individuals. However, past research on the structure of OSS developer networks has mostly neglected whether or not an individual was sponsored by a firm, and, more importantly, did not take into account the sponsoring firm's dedication towards OSS.[3]

In addition, considering the discussion on organizational ambidexterity (cf. Section 2.2.2), organizations are unable to simultaneously explore and exploit (O'Reilly III. & Tushman 2011). In a similar vein, individuals either focus on exploration (e.g., by communicating with peers in a network) or exploitation (e.g., by contributing to OSS code). Firms that assign developers to an OSSDP thus explore and exploit by means of their fellow employees. Consequently, affected by individual level attitudes and behaviors, firms transform their potential into realized absorptive and desorptive capacity on an organizational level (Lichtenthaler 2011c).

However, becoming a formal leader in open innovation communities requires technical contributions in advance (Fleming & Waguespack 2007). In contrast, informal authority is associated with coordination and communication work (Barley & Kunda 2001, Bechky

---

[2]By referring to actor-network theory, Tuomi (2001) emphasized that complicated sub-networks become represented by actants, and the complex underlying structure becomes a "black box" for practical purposes.

[3]See Dahlander & Wallin (2006) for one of the few exceptions.

2006, Romanelli 1991). If technical contributions are interpreted as exploitation and communication work as exploration, firm-sponsored developers have to dissolve a trade-off. By
delineating lateral from vertical lines of authority, Dahlander & O'Mahony (2011) portray how engagement in knowledge work differs according to technical problem solving
and coordination tasks. In particular, they found that receiving formal positions, such as
becoming a member in a board of directors, lead to an increase in coordination work while
limiting the amount of technical contributions. Conveyed to the firm as level of analysis,
this implies that once their fellow employees have reached formal authority, exploitation is
likely to decrease.

To sum up, communication intensity, network position, and technical contributions
within an OSSDP affect the likelihood of becoming a project leader. Whereas informal
or lateral authority is a function of an individual's network position, formal authority is
associated with technical contributions in advance. In their aim to control the project's
trajectory, firms – in general, and those with a business model dedicated to OSS in particular – seek to obtain central positions within the network of developers. Yet, there is a
dearth of research examining developer networks regarding an individual's network position, communication intensity, technical contribution, *and* sponsorship simultaneously.

To narrow this research gap, this section is devoted to the role of firm-sponsored individuals within a community-initiated MVP as a means to manage innovation beyond firm
boundaries and aims to further our understanding of how knowledge work is coordinated
in open innovation communities in general.

## 5.2   Hierarchy and Lateral Authority in Distributed Knowledge Work

It has been increasingly argued by organization theorists that in a postbureaucratic world
horizontal collaboration has often replaced vertical chains of command as a result of economic and organizational shifts, such as the unbounded access to information (e.g., Barley
1996, Barley & Kunda 2001, Kellogg, Orlikowski & Yates 2006, O'Mahony & Bechky
2006). Rather than pushing information back and forth through a coppice of hierarchical advice structures, in modern organizations work and knowledge exchange is based on
social ties and organized horizontally (Kotlarsky & Oshri 2005). Moreover, many tasks
are organized and executed at the project level. In addition, not only is work and knowledge sharing increasingly distributed within organizations, but across the boundaries of the

firm (Hansen 2002, Hansen et al. 2005). In that sense, OSSDPs mirror new organizational forms of distributed and collective work since multiple firms work collectively together and share knowledge across their boundaries especially in community-initiated MVPs (Koch & Schneider 2002).

However, with the shift from vertical to horizontal authority, new questions concerning knowledge sharing, power distribution, and career motives arise. For example, in hierarchical settings, vertical authority is considered to help leaders coordinate tasks, and, more importantly, to terminate debates (e.g., Dalton 1959). However, in the absence of hierarchical structures, authority is not concentrated in a single person but distributed among several actors (Doeringer & Piore 1971), a fact that slows down decision making processes (Benbunan-Fich, Hiltz & Turoff 2002). In addition, as noted by various career theorists, in the absence of hierarchical structures, career paths change significantly as there is no longer a career ladder to climb (Ibarra 2004).

By considering progression as gains in responsibility, Dahlander & O'Mahony (2011) argue that within non-hierarchical organizations, individuals progress towards the center (of the network they work in) instead of aspiring "higher" positions within a hierarchy. More precisely, because of their position, central persons are able to communicate more intensely and therefore receive authority over peers due to their ability to coordinate tasks, not because they were equipped as supervisors by a third party. Thus, individuals gain horizontal authority by achieving responsibility and decision rights for a greater proportion of the project rather than achieving responsibility over an increased number of subordinates.

In situations where hierarchical structures meet more horizontal structures, such with a matrix organization (Galbraith 1973), lateral authority might emerge in addition to vertical authority leading to possible conflicts in advice chains. To avoid ambiguity, Dahlander & O'Mahony (2011, p. 965) use the term lateral authority:

> as opposed to horizontal authority to indicate task-based authority, without necessarily inviting orthogonal contrast between horizontal and vertical systems. To operationalize lateral authority for empirical study, we define it as authority over collective work that does not include vertical authority over individuals.

Thus, people at the center of a community do not manage their peers, that is, they are not responsible for their career paths, training, and qualification – tasks typically to found on the management's agenda within firm boundaries. Instead, project management is accomplished by coordinating tasks only.

A precondition for lateral authority is what scholars refer to as participation inequality (Kuk 2006). Depending on the type of project, there is an imbalance between active and non-active participants. Various authors have used the onion-metaphor in order to describe participation inequality and to separate the core (of active developers) from the periphery (of occassionally participating bug reporters) (e.g., Lakhani 2006, MacCormack et al. 2010, Masmoudi et al. 2009). Figure 5.1 portrays an ideal core-periphery structure within an OSSDP.

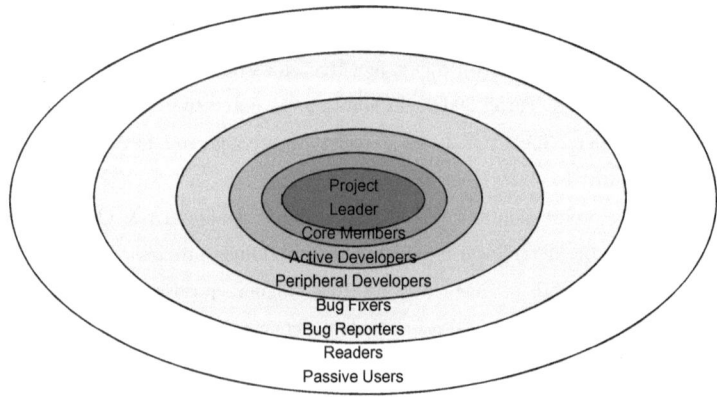

**Figure 5.1** The Onion Model of OSS Development (Source: Masmoudi et al. 2010, p. 289)

When all participants would contribute and communicate equally, no such horizontal authority structures would emerge. In particular, if any participant writes exactly the same number of code contributions in the same quality and undertakes the same effort in coordination work, no leader would obtrude. However, participation inequality is in stark contrast with the widely promoted bazaar model of OSS development (e.g., Demil & Lecocq 2006, Raymond 1998), which suggests that thousands of developers contribute equally to a project (Goldman & Gabriel 2005).

A self-evident strategy to obtain lateral authority within a non-hierarchical community would imply simply increasing one's level of activity in coordination work. Even if this effort would be valued by the community, which is not necessarily the case if the quality of work is worse, a person's ability to escalate his level of activity has natural limits. This gives other community members the chance to step in and attempt to gain responsibilities. This natural limitation of "coordinating capacity" is also mirrored by several studies on the structure of OSSDPs (e.g., Franck & Jungwirth 2003b, Long & Siau 2008, Setia, Rajagopal, Sambamurthy & Calantone 2011).

For example, Crowston & Howison (2006) show that the level of centralization (i.e. a few members write a great proportion of the code) within a project is negatively correlated with project size. Thus, larger projects become more modular. By considering coordination work as a form of knowledge sharing, Kuk (2006) highlights that participation inequality exhibits a curvilinear relationship with knowledge sharing. At the extremes, that is either all developers posted fairly evenly or only a few developers engaged in posting, knowledge sharing is less effectively than within moderate ranges of participation inequality. In a similar vein, based on simulation data for exploration and exploitation as complementary learning strategies in parallel problem solving, Lazer & Friedman (2007) found that for intermediate time frames, there is an inverted-U relationship between connectedness and performance in which both poorly and well-connected systems perform badly and moderately connected systems perform best.

It has been argued in the preceding chapter, that firms might obtain control over the project's trajectory by either assigning a large proportion of paid developers to the project (RDBC) or by capturing project leadership (CBL) (cf. section 3.4.4). The latter can take the form of formal and informal leadership (Aghion & Tirole 1997). For example, in the case of Eclipse, project leaders are elected by fellow developers. However, there are many other commercialized OSSDPs where leadership is either not allocated by election or does not even exist as a formal concept. Thus, in situations where de novo entrants seek to harness external development resources, achieving CBL by motivating firm-sponsored developers to progress to the center is a means to control (at least parts of) the project without being formally elected as a project leader.

## 5.3 Centralization and Contributions in Open Source Communities

Engagement in coordination work puts individuals in the position to receive lateral authority. For firms, enforcing their own developers to engage in coordination work, is thus crucial to control participants who are not employed by the focal firm and outside the firm's direct advice chain. This seems to be the most promising way to benefit from free external labor, which is considered by various researchers one of the main advantages of collaborating with external communities (e.g., Fey & Birkinshaw 2005, Mol & Wijnberg 2011, Walsh et al.

2011). However, following the extant literature pertaining to communities that are not limited by organizational boundaries nor by vertical authority, technical communication as well as technical contributions are equally said to be behaviors that favor progression to a lateral authority role (e.g., Dahlander & O'Mahony 2011, Fleming & Waguespack 2007, Okhuysen & Bechky 2009). All tasks, coordination work, technical communication, and technical contributions require specific knowledge. Whereas the latter two require advanced knowledge on technical aspects, coordination work requires knowledge about peers' skills. In addition, since in a distributed group with lateral authority only a few people actually coordinate, knowledge about developer skills is not shared among the group.

In contrast, technical communication as well as technical contributions enforce knowledge sharing, which has been shown to be one of the driving motivations to participate in innovation communities (e.g., Roberts et al. 2006, Shah 2006, Wu et al. 2007). From a knowledge sharing perspective, though, technical communication has to be distinguished from technical contribution (Hansen 2002, Wasko & Faraj 2000, Wasko & Faraj 2005). The latter is the result of a knowledge sharing process and therefore static (Hansen et al. 2005). Individuals are deprived of contributing their ideas since the outcome is not adaptable anymore. In contrast, when individuals communicate about technical challenges, they are stimulated by fun in problem solving and the prospect of being recognized for their solution. Thus, technical contributions refer to already solved problems and constitute a different kind of knowledge sharing than technical communication.[4]

Given that OSSDPs usually run on meritocracy, newly joined OSS developers are not allowed to contribute technical artifacts to the code base. In order to get recognized and to increase the chance of having their postings reciprocated, de novo entrants first have to climb the apprenticeship ladder (Kim 2000). What naturally emerges then is an informal division of labor within the community (e.g., Aiken & Hage 1968, Blau & Scott 1962, Hauschildt & Chakrabarti 1989) with new entrants concentrating on bug fixing and circumspectly participating in technical discussions and established developers contributing technical artifacts. In theory, though, one network based on coordination work, one based on technical communication, and one based on technical contributions exist. Depending on which network is chosen, investigations on power distribution and centrality might lead to different results (Ibarra 1993, Killduff & Tsai 2003).

---

[4]As will be shown in the research setting section, in this dissertation it is not possible to separate technical communication from coordination work as the data set used is content-agnostic.

When lateral authority is achieved by taking up central positions within a network, firms that intend to influence an OSSDP according to their interests may chose which kind of centrality they suggest their employees pursue. Yet little is known so far if firms enforce their own developers to achieve lateral authority by engaging in coordination work, technical discussions, or by contributing technical solutions (Fleming & Waguespack 2007, Henkel 2009). The following section therefore aims at suggesting firm behavior based on different theoretical anchors such as network theory, control theory, and the knowledge management capacity framework.

# 5.4 Hypotheses Development

Again, the hypotheses section in split into two parts. Part I pertains to the social structure of an OSSDP and the network position firm-sponsored individuals are likely to obtain. Seven hypotheses will be proposed that build upon social network theory as well as the concept of lateral authority. In Part II, the interplay of communication and technical contribution will be shifted in the focus.

## 5.4.1 Part I: Sponsorship and Lateral Authority

The following hypotheses 1 to 4 are lent from the pathbreaking work of Dahlander & Wallin (2006). However, a different argumentation line will be used to deduce behavior of firm-sponsored developers. For example, Dahlander & Wallin (2006) posit that hobbyists differ from firm-sponsored individuals according to their motivation to participate. Whereas fun in problem solving or signaling to a future employer are considered reasons for voluntary developers to participate in OSS development (e.g., Bitzer & Geishecker 2010, Hars & Ou 2002, Krishnamurthy 2006, Wu et al. 2007), firms are active in OSSDPs because they are interested in OSS as a complimentary product and because they gain access to resources they cannot buy on the market[5] (e.g., Lerner & Tirole 2002, Iansiti & Richards 2006). Firm-sponsored individuals, though, engage because they were told to do so.

Although these considerations are seemingly right, they are imperfect as they neglect what has been previously discussed in the light of knowledge management and lateral authority. In other words, the hypotheses have been introduced by referring to *why* individuals participate in OSS development, while disregarding *how* individuals differ in their

---

[5]Dahlander & Wallin (2006) argue that no market exists where firms can "buy" free external resources. However, it is possible to internalize these external resources by simply offering them a job.

engagement in terms of knowledge sharing and technical activity. Dahlander & Wallin (2006) have further noted that firms are more active than volunteers primarily for two reasons, namely because they *can* as a result of their resource advantage and because they *have to* in order to prevent being considered a free rider. Regarding the enormous investments that have to be made by firms that engage in OSS development (i.e. paying developers and foundation membership fees), I must add *because they want to* to this reasoning. More precisely, firms that have a commercial interest in the OSSDP, for instance, by pursuing a business model dependent on a particular OSS, engage in OSS development because they intend to control the project.

As discussed, in the absence of hierarchical control structures, firms might apply what has been defined as RDBC and CBL. The latter may be achieved by either being elected as a formal leader or by progressing to positions of lateral authority. Since the leadership role typically devolves upon the individual in the most central position (Leavitt 1951), individuals sponsored by a firm with a business model dependent on OSS should be more central than those sponsored by firms without an OSS business model. In turn, both types of sponsoring should lead to more central positions than being active as a volunteer.

*Hypothesis 1:* Being sponsored by a firm with a business model dedicated to OSS increases the likelihood that an individual seeks to form ties with other participants in the community.

Centrality, however, can be calculated in various ways (Wassermann & Faust 1994). For example, in bidirectional network relations, such as in the case of communication within an OSSDP, initiating relations to peers differs from responding to communication requests. Whereas knowledge seekers tend to ask for knowledge by initiating communication, knowledgeable individuals, in contrast, respond to requests. Because of reciprocity and the gift exchange culture within OSSDPs (Bergquist & Ljungberg 2001, Walsh et al. 2011), initiating communications should be highly correlated with responding to communication requests.[6] Thus, if firm-sponsored individuals are likely to seek to form ties, they are also likely to receive connection requests. In addition, due to their resource advantage, firm-sponsored developers are more knowledgeable and constitute a target for knowledge seekers within the community.

---

[6]As noted by Dahlander & O'Mahony (2011), with regard to knowledge sharing, technical communication has to be distinguished from coordination work. Since the data used is content-agnostic and therefore prevents deeper analysis, I skip a discussion on the differences here.

*Hypothesis 2:* Being sponsored by a firm with a business model dedicated to OSS increases the likelihood that participants in the community seek to form ties with the individual.

Individuals who are sought after within a community are said to be prestigious. They are well-known and usually possess superior knowledge that increases the likelihood of being considered a communication partner. Dahlander & Wallin (2006) argue that firm-sponsored individuals are less prestigious[7] since they have to legitimize their commercial endeavors by exchanging intensely with the community – despite their knowledge advantage.

*Hypothesis 3:* Being sponsored by a firm with a business model dedicated to OSS decreases the likelihood that an individual is prestigious in the community.

Having connections to a large number of peers favors an individual's ability to absorb specific knowledge. However, it is obvious and has been highlighted by various social network researchers (e.g., Bonacich 1972a) that having a high number of ties alone might not be enough in order to gain central network positions, not to say knowledge advantages. In fact, being connected to knowledgeable ties is more advantageous than being connected to a high number of peers (Faust 1997, Rost 2011). Individuals are knowledgeable because they are well connected. Thus, firm-sponsored individuals seek to interact with well-connected individuals, who, in turn, are likely to be firm-sponsored themselves.[8]

*Hypothesis 4:* Being sponsored by a firm with a business model dedicated to OSS increases the likelihood that an individual is tied to well-connected participants in the community.

Hypotheses 1-4 all refer to network centrality that is generally associated with power (e.g., Bonacich 1987), knowledge advantage on the individual and firm level (e.g., Hansen 1999, Uzzi 1997), and lateral authority (e.g., Dahlander & O'Mahony 2011). However, given the pluralism of centrality measures (cf. Freeman 1979, Wasserman & Faust 1994), it is important to (1) test for other forms of centrality and embeddedness and (2) to test for effects more deeply.

For example, based on Granovetter's (1973, 1983) distinction of strong and weak ties, supporters of the theory of structural holes highlight that opinions and behaviors within a

---

[7]Within social network theory, prestige is defined as the difference between being the receiver and being the initiator of communication requests. See 5.5.3.1 for further details.

[8]The notion that firm-sponsored developers are more connected to other firm-sponsored developers than to other peers reflects both clan control and the R&D alliance character of commercial OSSDPs.

group are more homogeneous than between groups (Burt 1992). Consequently, people who are connected across groups are therefore more familiar with alternative beliefs and ways of thinking (Burt 2004). Since diversity in thinking is considered a driver for creativity, those people are likely to produce "better" ideas than their peers (Amabile et al. 1996). Thus, individuals with the same number of direct ties do not necessarily constitute a homogeneous group of "central individuals" as suggested by centrality measures such as degree centrality. Moreover, it is possible that an actor is only connected to peers in a subgroup of a network rather than being connected to peers from other subgroups. In Figure 5.2, for example, actor A and actor D possess an identical number of direct ties, but A is in an advantageous position due to his access to more diverse opinions and behaviors (Kleinberg, Suri, Tardos & Wexler 2008).

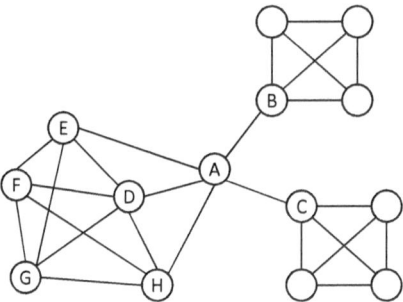

**Figure 5.2** Structural Advantage and Structural Holes in Social Networks

People at the edge of a social network that nurtures connections mainly with people outside their network are referred to as boundary spanners (Cross & Prusak 2002). The term stems from situations within structured organizations and is used to describe people who are able to connect distinct organizational units, such as marketing with accounting. Moreover, those people even maintain connections to parties that are located outside the boundaries of the firm, such as universities. In a similar vein, Kuk (2006, p. 1034) emphasizes that:

> strong structural ties are said to enhance transfer of complex intangible resources including knowhow and embedded knowledge [...] through improved communication as members often find themselves engaging in similar and overlapping activities [...].

However, despite the knowledge advantages of being accepted by multiple groups, boundary spanners also face various challenges, such as role ambiguity and role conflict (Friedman & Podolny 1992, Van Sell, Brief & Schuler 1981).

Similar to boundary spanners, information brokers are characterized by a wealth of indirect connections, which makes them even more important than actors with a higher number of direct ties (Long Lingo & O'Mahony 2010). Cross & Prusak (2002) distinguish boundary spanners from information brokers by the domain in which they are active: Whereas information brokers only act within the social network, boundary spanners connect different social networks. Accordingly, this distinction was based on the existence of organizational borders. In the case of OSS, both roles coalesce.

In summary, boundary spanners as well as information brokers feature knowledge advantages over their peers. Literature pertaining to structural holes and social capital underpins that if those people retire from their engagement, they leave structural holes that slow down the network's performance (e.g., Armbruster 2005, Davenport & Daellenbach 2011, Burt 2007, Méndez-Durón & Garcia 2009, Nahapiet & Ghoshal 1998). Thus, individuals seeking to gain central positions may quickly use accrued structural holes to benefit from existing relations rather than build all necessary connections independently. Since "brokers less likely have ideas dismissed and [are] more likely to have ideas evaluated as valuable" (Burt 2004, p. 349), firms should be likely to gain brokerage positions in order to push their agenda. Building on the above reasoning, brokerage positions are thus a more appropriate way to gain lateral authority than more simple forms of centrality, such as just being connected to more peers (Fleming & Waguespack 2007). Hypotheses 5 and 6 are devoted to these aspects.

*Hypothesis 5:* Being sponsored by a firm with a business model dedicated to OSS increases the likelihood that an individual leaves structural holes if he retires from the project.

*Hypothesis 6:* Being sponsored by a firm with a business model dedicated to OSS increases the likelihood that an individual seeks to achieve brokerage positions.

Finally, although firm-sponsored individuals exhibit knowledge advantages compared to their voluntary counterparts, their employing firms are still interested in absorbing external knowledge that complements the internal knowledge resource base (Jensen, Johnson, Lorenz & Lundvall 2007, Lichtenthaler & Lichtenthaler 2010). This implies that even if firms do not nurture connections to peripheral specialists intensely, it is still important to be able to contact those people quickly if necessary. For example, Aral, Brynjolfsson & Van Alstyne (2007) have analyzed e-mail writing behavior and information flows within a job agency and report that path length reduces the likelihood of receiving information, with each additional hop reducing the likelihood of diffusion by 29%. In addition, they found support for the notion that receiving information earlier has a positive impact on individual productivity. In line with these findings, firm-sponsored individuals should be likely to seek positions with short communication paths.

*Hypothesis 7:* Being sponsored by a firm with a business model dedicated to OSS increases the likelihood that an individual seek to achieve positions with short average communication distance to other participants in the community.

### 5.4.2   Part II: Sponsorship and Technical Contributions

Dahlander & Wallin (2006) showed that individuals sponsored by a firm communicate more intensely than their voluntary peers. In extended research, Dahlander & O'Mahony (2011) revealed that the likelihood of becoming a leader in an OSS project is related to the intensity of prior communication, in both task execution and organization building. But once they reached leadership positions, their communication intensity decreases. In addition, O'Mahony & Ferraro (2007, p. 1092), who provided similar results on the emergence of governance, stated "from our qualitative data, it would appear that developers engaged in organization building would be more likely to assume leadership than those who were more hands-off or more concerned with technical issues."

Yet few studies have been able to combine aspects of sponsorship, communication intensity, and technical contributions. Several questions have been left unanswered: If firm-sponsored developers communicate more intensely in order to gain lateral authority, do they simultaneously take part in the development of technical contributions? In addition, the results of O'Mahony & Ferraro (2007) suggest that people either engage in managing the community or in technical development. However, does this hold true for firm-sponsored developers as well?

For de novo entrants, receiving lateral authority by progressing to the center is a suitable strategy to increase the level of control over an OSSDP (Dahlander & O'Mahony 2011). However, firm-sponsored developers who have been active in the project for a long time already occupy positions of authority. In addition, due to their merits about the project they have already gained positions where they are allowed to contribute to the code directly. Contributing, however, is more effective than (indirectly) controlling peers since the firm's conception can directly be integrated into the source code. In the end, firms are interested in a technical solution that aligns with their business model rather than being valued by a community for their engagement. In other words, achieving lateral authority is important if the firm is unable to control the code or external developers directly. In the case in which the firm possesses the right to contribute to the source code, it will be likely to engage in technical development. Since contributing, though, may be considered the highest form of project control, firms with a business model dedicated to OSS are even more likely to be responsible for technical contributions than firms without an explicit OSS business model.

*Hypothesis 8:* Being sponsored by a firm with a business model dedicated to OSS increases the likelihood that an individual is responsible for more technical contributions than volunteers.

Literature pertaining to organizational ambidexterity has taught us that firms face difficulties simultaneously exploring and exploiting with the same intensity (He & Wong 2004, O'Reilly III. & Tushman 2011, Raisch & Birkinshaw 2008). At the individual level,[9] even the best people can hardly effectively communicate intensely and contribute high quality solutions at the same time.

From a knowledge management perspective, individuals who seek to form ties with other individuals are searching for knowledge. Especially in mailing lists, those who initiate discussions usually lack necessary knowledge to pursue a task. Thus, by initiating new discussions, they are hoping to close their knowledge gaps. In turn, individuals who feature (knowledge) resource advantages are more likely to be addressees of knowledge seekers.[10] However, if seeking to form ties with other individuals in the community signals a lack of knowledge, than those people are likely to pursue less complex tasks such as bug reporting. Since their knowledge gap excludes them from contributing to the source code, they will not engage in technical contributions. In turn, those who are knowledgeable enough to contribute source code to the project are less likely to search for knowledge in the community first because they already possess the necessary knowledge to solve complex problems and

---

[9] According to Lichtenthaler (2011c), organizational capabilities such as absorptive and desorptive capacity are a direct consequence of individual capacities on task and project levels.

[10] Firm-sponsored developers do have resource advantages and can devote their entire time to the project, while volunteers cannot because they have to work in their real life jobs. Indeed, as shown by various researchers (e.g., Lakhani & Wolf 2005), firm-sponsored developers spend significantly more hours per week on their projects than volunteers.

second because they dedicate their time to programming rather than to communication. In summary, it can be posited:

> *Hypothesis 9:* Seeking to form ties with other individuals in the community decreases the likelihood to be responsible for technical contributions.

As discussed, being able to contribute source code to the repository enables developers to control the project's trajectory directly. Ceteris paribus, this would imply that once an individual has gained the right to contribute, he could discontinue his communication work and solely control by contribution. However, OSS communities have a long tradition of valuing openness and technical discussions. Thus, ignoring communication within the community will inevitably lead to an annoyed community and will make it difficult to maintain positions of authority in the long run. Individuals with advanced knowledge, though, are being sought by participants within the community and do retain from seeking to form ties themselves in favor of an increased number of technical contributions.

> *Hypothesis 10:* Being sought by participants to form ties with them increases the likelihood to be responsible for technical contributions.

## 5.5 Research Design

I chose a successful large OSS community to analyze the role of individual developers who are employed by participating firms. In general, a community around an OSS project may consist of thousands of participants who contribute unequally (Kuk 2006). Consequently, various researchers have found out that core-periphery structures are predominant in communities surrounding an OSSDP (e.g., Lakhani 2006, MacCormack et al. 2010, Masmoudi et al. 2009). Accordingly, few individuals may be called contributors and regularly contribute pieces of software code. Rather, the majority of participants report bugs or request new functionalities occasionally. In this sense, the developer network has a centralized and hierarchical structure whereas the user network is more decentralized (Crowston & Howison 2006).

Regarding the developed hypotheses, I am interested in the network position of firm sponsored developers as well as their technical contributions. To test the formulated hypotheses, data about an OSS community is needed that was gained by a meticulous data retrieving process. Thus, the aim of this section is to depict the objective of the research, to retell the process of data gathering, and to introduce the required variables.

## 5.5.1   Research Objective: The Linux Kernel Project

In the present study, I focus on the Linux kernel project. Since people often use Linux when they actually mean the Linux kernel, it is important to distinguish Linux from the Linux kernel. Linux is a UNIX based[11] open source operating system that exists in various derivative forms. The Linux kernel constitutes the core which is responsible for basic functions (e.g., hardware control). Thus, the Linux kernel can be considered the basis for any Linux distribution, such as Debian, Fedora or Ubuntu (e.g., Bovet & Cesati 2002).

In order to outline recent developments in Linux systems, Andreas Lundqvist and Donjan Rodic maintain a website, which is designated to delineate the timeline of different main Linux distributions and their derivatives. Figure 5.3 shows exemplarily the timeline of major RedHatLinux derivations since 1994.

Additionally, the Linux kernel is the core of many derivatives such as mobile operating systems (e.g., Android) or embedded devices (e.g., machine controls) (Gruber & Henkel 2006, Hampe 2010, Henkel 2007). Whereas the Linux kernel is developed by a group of distributed individual programmers, distributions may be produced within the boundaries of the firm as long as the final product aligns with the project's software license.[12]

The Linux kernel project was started in 1991 by Linus Torvalds at the University of Helsinki. In the beginning, not only did Torvalds share the code with interested people but he organized requests and the work of programmers who freely devoted their time to the project. Owed to the speedy pervasion of Linux and to the increasing number of people who voluntarily contributed to the project, Torvalds had to accept that interaction with each developer got too complex. Thus, as a natural response, a governance approach emerged that was based on a predefined role concept. *Maintainers* are responsible for particular modules of the kernel or for stable elder versions (Moon & Sproull 2002, Shaikh & Cornford 2003). *Credited developers*, sometimes referred to as *trusted lieutenants* (Dafermos 2001, Iannacci 2005), are those persons who contributed a substantial amount and quality of software code to the kernel (Moon & Sproull 2002). In addition, as in other successful communities, a core-periphery structure is observable (Lee & Cole 2003), which, in the case of Linux, may be considered a two-stage hierarchy (Cornford et al. 2010).

---

[11]It is important to note that although Linux's architecture is oriented toward UNIX, Linux does not contain any UNIX code.

[12]Currently, the Linux kernel is distributed under the GPLv2 license. However, recent developments made it necessary to implement proprietary components, such as the machine language BLOB. Therefore, in principle two versions of the Linux kernel exist, which are not 100% equal because for the OSS version the proprietary components were removed. For this study, I used the Linux kernel under the GPLv2.

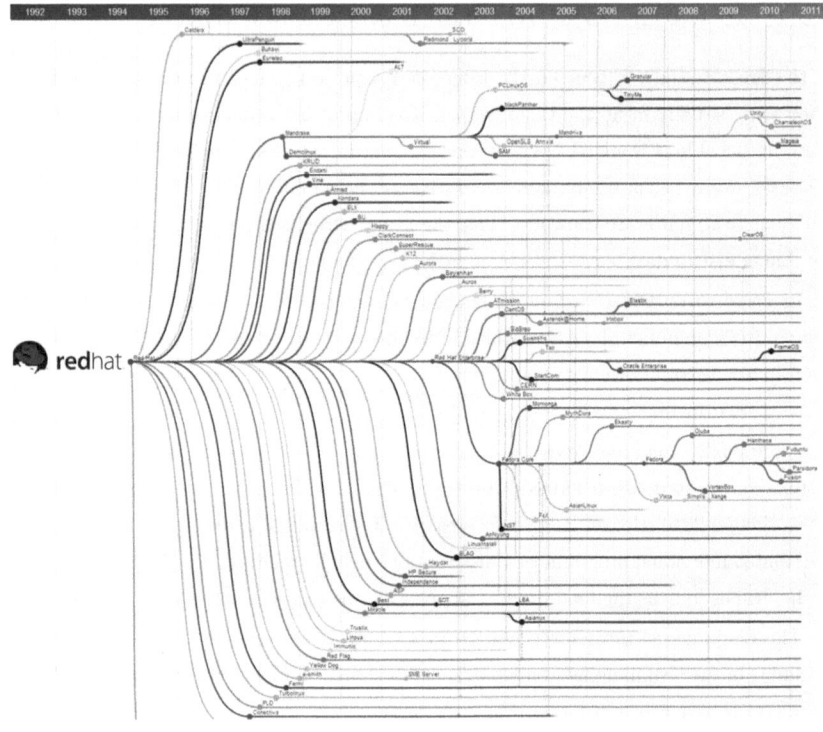

**Figure 5.3** RedHat Distribution Timeline (Source: futurist.se/gldt)

In 2007, the Linux foundation was founded as a legal body for the Linux kernel project. In contrast to other foundations (cf. Eclipse foundation in Chapter 4) foundation membership is distinguished from project membership. Foundation membership fees for firms range from US $5,000 to US $500,000. The aim of the non-profit organization is to promote, protect, and standardize Linux. These tasks entail marketing of Linux, protecting the IP of developers, and maintaining public relations (De Laat 2007).

The Linux kernel project is markedly active and successful. The community consists of more than 6100 individual developers and over 600 participating companies (Corbet, Kroah-Hartmann & McPherson 2010, Kroah-Hartmann, Corbet & McPherson 2008). Between 2007 and 2008, on a daily average, 4300 lines were added, 1800 lines were removed, and 1500 lines were modified. This is 3.69 changes per hour. In sum, today Linux consists of more than 13 million lines of code.[13] The growth of Linux is delineated in Figure 5.4. Despite the relevance of Linux for various firms and their business models, it is surprising that between stable release 2.6.30 and 2.6.35 approximately 19% of all changes to the source code still were made by people without an affiliation to any company involved in the development of Linux. However, the majority of developers are paid for their work on the kernel. Top contributors between versions 2.6.30 and 2.6.35 were RedHat (12%), Intel (7.8%), Novell (5%), and IBM (4.8%). Firms active in the market for mobile operating systems contributed comparatively little to the Linux kernel with, for instance, 0.7% (Google), 2.3% (Nokia), and 0.6% (Samsung) (Corbet et al. 2010).

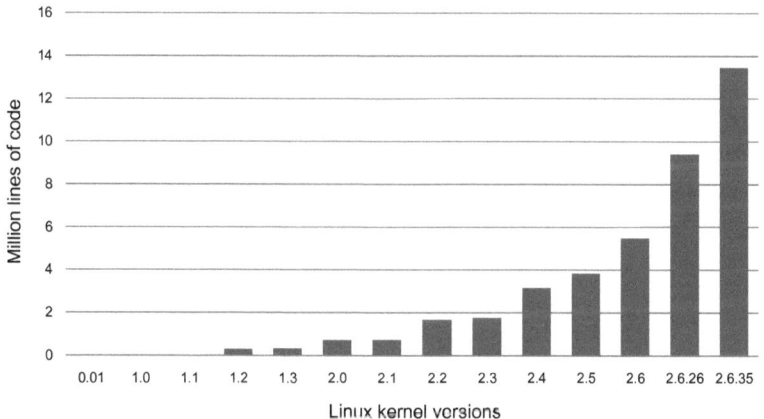

**Figure 5.4** Linux Kernel Source Code Size (Figures taken from Corbet et al. 2010)

Developers primarily communicate via mailing lists, that is, they send an e-mail to an e-mail distribution system. In order to receive messages posted to that mailing list, developers have to take out a subscription to that mailing list. In turn, if someone writes an e-mail it is distributed to all mailing list subscribers. Mailing lists are used for various purposes, such as announcements or requests for help (Dahlander & Wallin 2006). Lee & Cole (2003, p. 637) describe the mailing list's importance as follows: "The development work takes place mainly at the Linux kernel mailing list, which is a virtual environment where Linux developers send their contributions, discuss implementation details, and interact with other

---

[13]Ryan, P.: Linux kernel: 13 million lines, over 5 patches per hour. URL, http://arstechnica.com/opensource/news/2010/12/linux-kernel-13-million-lines-over-5-patches-per-hour.ars, last access 5/19/2011

developers." Despite being able to post messages to the mailing list, most subscribers are passive readers rather than active communicators. It is important to note that if someone replies to a message, this message is linked to the original one. All messages that refer to the same original message constitute a *thread*. Although any subscriber is able to read all messages within a thread, only those who send messages to a thread are considered to actually *use* the mailing list according to its intent.

Beside the main mailing list for Linux kernel developers, various sub mailing lists exist that are devoted to sub systems of the kernel (Hertel, Niedner & Herrmann 2003). All mailing lists are filed on web servers that provide free access even for non-subscribers. The web archive *marc.info* was chosen due to its parser friendly structure (cf., Schneider 2009).

## 5.5.2    The Process of Data Gathering

Testing the hypotheses requires drawing a social network of the developers engaged in Linux development. For this study, e-mail data was used since it provides advantages over survey self reports in that measures are objective and not biased by reconstruction of individuals (Aral, Brynjolfsson & Van Alstyne 2012, Marsden 1990, Reagans & McEvily 2003). In order to extract data from mailing lists and to preprocess it for the use with statistical programs, a software package that consists of four independent modules was developed.[14] The first module of the software package is a parser, that stores data from the *marc.info* server, a server that hosts the entire Linux kernel mailing list, in a MySQL database.

The second module of the software package analyzes information for each person within the database and generates a *mapping table*, which maps multiple e-mail addresses to a single identity. The third module is responsible for parsing the source code, namely the Linux kernel, and to map those who participated in the development of the code to those who wrote e-mails. Finally, module four takes the data stored in the MySQL database and transforms it to a format that allows statistical programs to calculate predefined measures. In the following, I will briefly report how each part works in detail.[15]

**Mailing list parser.** Parsing is known as computer supported syntactic analysis. More precisely, a parser refers to a computer program that is able to fragment strings of text

---

[14]The development of the software was outsourced to the master thesis of Christoph Schneider, University of Koblenz-Landau (see Schneider 2009). I can't thank him enough for his effort. He did a great job and without his work, this thesis would definitely have to look differently.

[15]Those who are interested in a more detailed description may have a look at Christoph Schneider's master thesis (Schneider 2009, pp. 47-70).

inputs. Parsing is necessary to transform the information on a website (e.g., name of a developer) in a format that can be stored in a relational database. The mailing list parser consists of two independent modules, a *StructureParser* and a *ContentParser*. This separation is a prerequisite for successful parsing as the content parser has to know about the structure of the website in advance. Both modules follow a model-view-controller architecture (cf., Gamma, Helm, Johnson & Vlissides 1995). The *StructureParser* consists of a *MonthStripper*, a *ThreadStripper*, and a *MessageStripper*. All strippers operate in a sequential processing. First, the *Monthstripper* delivers information about the number of months a single mailing list was active. Second, the *ThreadStripper* identifies every thread within a month. In particular, the *ThreadStripper* extracts the number of messages within a thread and turns the URL over to the *MessageStripper*. Figure 5.5 shows 30 threads on a page within a mailing list month. "Next" and "Prev" are used to navigate to other months. With this procedure, the structure of every mailing list concerning the Linux kernel in analyzed on a monthly basis. Third, the *MessageStripper* extracts every URL

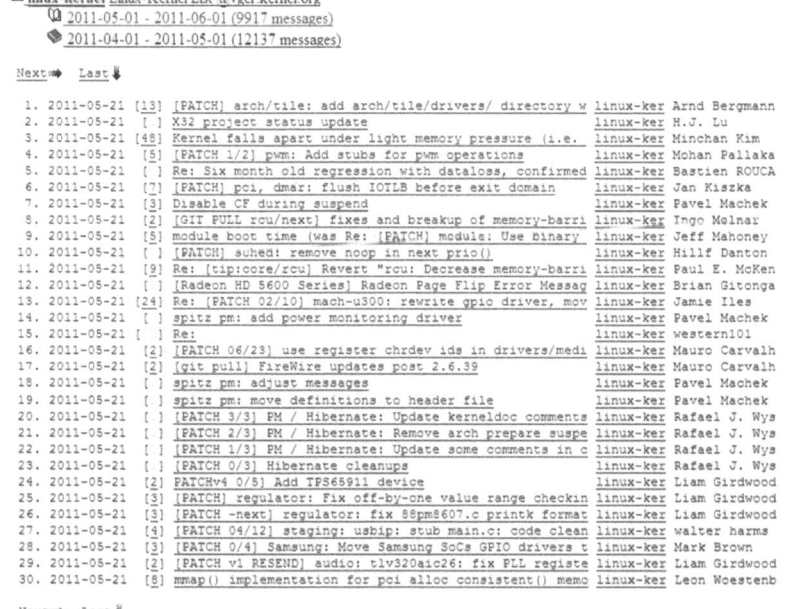

**Figure 5.5** Overview of Threads Within a Month (Source: marc.info)

of each message and stores it in an xml-format. Figure 5.6 shows the assembling of an
xml-structure file.

```
1   <?xml version="1.0" encoding="UTF-8"?>
2   <urlList>
3     <month name="2003-01-01_-_2003-02-01" url="http://63.238.77.172/?l=linux-kernel&
        amp;r=1&amp;b=200301&amp;w=2">
4       <thread page="1" name="linux-kernel,_you'll_like_that_jkpxton" url="http:
          //63.238.77.172/?l=linux-kernel&amp;m=104405755924934&amp;w=2">
5         <message name="linux-kernel,_you'll_like_that_jkpxton" url="http:
            //63.238.77.172/?l=linux-kernel&amp;m=104405755924934&amp;w=2" />
6       </thread>
7       <thread page="1" name="Re:_[Bitkeeper-announce]_Re:_bkbits.net_downtime" url="
          http://63.238.77.172/?t=104405333100001&amp;r=1&amp;w=2&amp;n=3">
8         <message name="Re:_[Bitkeeper-announce]_Re:_bkbits.net_downtime" url="http:
            //63.238.77.172/?l=linux-kernel&amp;m=104405457821654&amp;w=2" />
9         <message name="[Bitkeeper-announce]_Re:_bkbits.net_downtime" url="http:
            //63.238.77.172/?l=bitkeeper-announce&amp;m=104405316119852&amp;w=2" />
10      </thread>
11      ...
12    </month>
13  </urlList>
```

**Figure 5.6** Structure of an XML-file Design to Represent a Mailing List's Structure

The configuration of the *ContentParser* is more complex. The *ContentParser* module is
programmed parallel. This is owed to the enormous amount of data the parser has to
handle. For example, all Linux kernel mailing lists together consist of approximately 1.3
million messages for the period from 2000-2008. The *ContentParser* operates in four major
steps. In step one, the parser calls the xml-structure-file for each mailing list. After a test
of existence – the URL might not be reachable due to bad server connections – in step two
the HTML page is called whose URL was stored in the xml-file. In step three, the HTML
page is fragmented. Due to various "request for comments" (RFC), that is, accepted quasi
standards for computer network engineering published by the Internet Engineering Task
Force (IETF), e-mails in developer mailing lists follow a standardized structure. The first
17 lines of an HTML page embrace information about, author, topic, time, the sender's
e-mail address, and addressee. The content starts with line 18. An example cutout of a
message is depicted in Figure 5.7.

Because of the standardization, it is further possible to separate information about
a message from its content. After the parser has cut off the content of the message,
the content is stored in the database. Regarding the "header" of the message, storing is
more challenging as information would get lost if the header would be stored as a string.
Another sub-procedure, the *MessageParser*, searches for key words, such as "subject" or

```
[prev in list] [next in list] [prev in thread] [next in thread]

List:        linux-kernel
Subject:     Re: [git pull] FireWire updates post 2.6.39
From:        Mauro Carvalho Chehab <mchehab () infradead ! org>
Date:        2011-05-21 12:43:32
Message-ID:  4DD7B374.7040907 () infradead ! org
[Download message RAW]

Em 21-05-2011 07:07, Stefan Richter escreveu:
> Linus, please pull from the for-linus branch at
>
>      git://git.kernel.org/pub/scm/linux/kernel/git/ieee1394/linux1394-2.6.git for-linus
>
> to receive the following updates for the IEEE 1394 (FireWire) subsystem.
> They contain small fixes and some optimizations, e.g. to CPU utilization
> during isochronous I/O.
>
> The changes in (drivers/media/dvb,sound)/firewire/ have been Cc'd to
> linux-media and alsa-devel but I did not request explicit Acks.  There
> are no conflicts yet and are unlikely to arise during the merge window.
>
>   drivers/media/dvb/firewire/firedtv-avc.c  |   15 +------
>   drivers/media/dvb/firewire/firedtv-fw.c   |    1 +

The changes on the above dvb stuff are due to the API changes and looks fine
for me.

Acked-by: Mauro Carvalho Chehab <mchehab@redhat.com>
--
To unsubscribe from this list: send the line "unsubscribe linux-kernel" in
the body of a message to majordomo@vger.kernel.org
More majordomo info at  http://vger.kernel.org/majordomo-info.html
Please read the FAQ at  http://www.tux.org/lkml/
[prev in list] [next in list] [prev in thread] [next in thread]
```

**Figure 5.7** Cutout of a Message (Source: marc.info)

"from", and fragments the entire string in substrings according to a logical structure. For example, the key word "from" is always followed by a string that contains information about an author's name and e-mail address; "date" is always followed by the date and time the message was written. Finally, in step four, the substrings get stored in the MySQL database according to their attributes, such as person (data, which refers to a person), thread (data, which refers to a thread), message (meta-data of the message), and content (content of the message). In order to identify relationships between individuals, the parser first checks if there is a dedicated addressee. If there is no such addressee, the parser draws relationships to each individual who has ever posted in the focal thread.

**Mapping table.** Generally, e-mail data is considered to be inherently disadvantageous for several reasons. First, persons may use different e-mail addresses, such as their corporate address or a private one. Second, e-mail addresses as well as affiliations may change over time. Third, sometimes people even use different names with the same e-mail address. Taking the representation of all Linux kernel mailing lists stored in the database "as is" implies accepting a bad quality of the dataset. To overcome the data quality problem, the concept of a *mapping table* was developed. This table consists of only two columns and is generated by a separate module of the parser. Column one contains an unique identifier $pid_1$ of a person $p_1$. The second column contains a second identifier $pid_2$. The parser identifies both persons $p_1$ and $p_2$ as individuals, but in reality it is the same person who used either different e-mail addresses or used different names with the same e-mail address. Deciding if two *pids* refer to the same person manually is impossible given a dataset of more than 100.000 *pids*. Thus, an algorithm was developed to execute this task.

In a first step, the algorithm searches for *pids* with identical entries in the name field. As people who contribute to the Linux development and who respect RFCs mention their full name, the program checks for space characters that separate first from last name and for entries with at least six characters. By using this method, unrealistic names, such as *test* are excluded from the analysis. Additionally, the program checks if a name is listed on a black list. The black list consists of common senders of e-mails that are not individuals (e.g., Mail Delivery Service). After ignoring unrealistic senders, the algorithm maps persons with different *pids* but with the same name to a single *pid*. Admittedly, two different persons still may indeed have the same name, but as this is very seldom,[16] this limitation was accepted.

In a second step, the algorithm searches for *pids* with identical entries in the e-mail field. As in the first step, the algorithm searches for key characters in the text string. For e-mail addresses, one specific sign is the "@" character as well as a dot in front of the country postfix. In addition, the algorithm searches for entries in a second blacklist. These list entries encompass automatically generated e-mail accounts as well as senders addresses who do not refer to individuals, such as *mailer-daemon*. The algorithm is therefore able to map several database entries to a single identity regardless of the number of e-mail addresses a person has used.

**Source code parser.** The source code parser is designed to traverse every source code file under each root directory. Similar to the structure of e-mails, each source code file follows a standardized structure. For example, in the copyright section, all authors who contributed to the file are mentioned (see Lee & Cole 2003, Moon & Sproull 2002). The

---

[16]Checking the first 600 entries of the data base manually revealed that no combination of a person's first and last name occurred twice.

core of the source code parser is a function that compares name and e-mail database entries with their appearance in the source file. In particular, the function takes a single source file and checks for every identity if e-mail address or name of the person occurs in the source file. This comparison is very computationally intensive. Approximately 11.000 source files have to be cross checked with more than 100.000 database entries, which requires more than $1.17 \cdot 10^9$ calculations. In a last step, the parser investigates if developers who are listed in the MAINTAINERS or CREDITS file, respectively, have a *pid* in the database in order to identify developers with specific roles within the project.

**Adjacency matrix creator.** To calculate measures for social network analysis, software packages such as UCINET for Windows (Borgatti et al. 2002) or Pajek (De Nooy, Mrvar & Batagelji 2005) are used. In order to align with the format requirements for these programs, the information in the database has to be extracted and transformed in so-called adjacency matrices. The *AjazenzMatrixCreator* delineates the *AdjStructureRetriever* and the *AdjListCreator*. The *AdjStructureRetriever* works on a monthly basis. For each month, it eliminates redundancies (e.g., if a message was posted to numerous mailing lists). Subsequently, relationships between messages and persons are drawn. By iterating over the message list, the *AdjListCreator* checks if a message had a specific addressee. If no specific addressee was identified or in case the message was intended for every mailing list recipient, the function draws a relationship between the sender and each person who posted at least once in the focal thread. Afterwards, each person who is an integral part of an edge is analyzed with regard to affiliation, number of co-developed source files, and appearance in the MAINTAINERS and CREDITS file. As affiliation may change over time, the function additionally counts the occurrence of e-mail postfixes for each person.

The network structure of the Linux community is represented by means of an algebraic graph. In algebraic graph theory, a graph consists of nodes and edges (e.g., Biggs 1993, Borgatti & Everett 2006, Krackhart 1993, Krumke & Noltemeier 2005). While developers are symbolized by nodes, e-mail exchange between individuals is symbolized by edges from node to node. Thus, the prerequisites for drawing a complete network of relationships between developers are given. Finally, the *AdjListCreator* creates a matrix that represents the network structure. In detail, '0' represents that there is no relationship between two individuals; '1' represents that a relationship exists. These matrices are taken to calculate specific measures concerning the network structure of the Linux developer community (e.g., Wassermann & Faust 1994).

In summary, the developed software package is able to transform rather unstructured website data into a format that enables statistical calculations. In particular, by using this laborious procedure as well as the UCINET and Pajek software packages, all required variables can be constructed out of the available data.

### 5.5.3   Operationalization of Variables

Using the mailing list data and the source code data I obtained, it was possible to calculate various measures concerning individuals' communication and programming behavior. In particular, I used data from 2006-2008 as in 2006 the major release Linux kernel 2.6 was introduced. The measures for network position are based on graph theory (Krumke & Noltemeier 2005) as well as social network theory (Faust 1997) and were calculated using UCINET for Windows (Borgatti et al. 2002) and Pajek (De Nooy et al. 2005).

#### 5.5.3.1   Dependent Variables

As pointed out in previous studies on firms in OSS development, firm-sponsored individuals are more likely to obtain central positions in a network (e.g., Dahlander & Wallin 2006). Following this assumption, I calculated classical measures for network position, such as actor centrality (e.g., Bonacich 1987, Faust 1997, Freeman 1979, Wassermann & Faust 1994) and more uncommon ones, such as constraint centrality (Burt 1992, De Nooy et al. 2005). Central to this analysis is that I am able to distinguish if an individual has sent or received an e-mail. Thus, the analysis is based on directionality. Since for some network measures several terms exist that are confusingly used interchangeably, I will discuss and compare different notions when needed.

**DEG-CENTRALITY.** *Degree centrality* is a rather simple measure and is calculated by the number of actors an individual has sent an e-mail to. If $n$ is the number of actors within the network, this measure may have values ranging from 0 to $n-1$. As the e-mails are sent from the individual to others, some researchers refer to Degree centrality as *Outdegree* (e.g., Dahlander & Wallin 2006).

**DEG-PRESTIGE.** *Degree prestige* may be considered as a counterpart to degree centrality. It describes the number of persons who initiated a communication towards the individual. The measure equally may take values from 0 to $n-1$. Consequently, in Dahlander

& Wallin's (2006) work, this measure is referred to as *Indegee*. In other contexts, this measure is used as an index for *popularity* (e.g., De Nooy et al. 2005).

**PRESTIGE.** *Prestige* indicates to what extent other actors have initiated more communication to the focal actor, than the focal actor has initiated to other peers (Wassermann & Faust 1994). If $n$ is the size of the network (i.e. number of participating actors), prestige may have values between $-(n-1)$ und $(n-1)$.

**EIG-CENTRALITY.** *Eigenvector centrality* is an index that not only represents with how many actors an individual has relations, but which takes into account the importance of the actor an individual has a relation to (e.g., Bonacich 2007, Grewal et al. 2006). More precisely, EIG-CENTRALITY is a measure in which a unit's centrality is its summed connections to others, weighted by their centralities (Bonacich 1972a, Bonacich 1972b). Values lie between 0 and 1.

**AGG-CONSTRAINT.** *Aggregate constraint* is a measure that is used rarely but is of high interest to control and brokerage. It specifies to which extant an actor P is in a more beneficial relation to other actors than the other actors among themselves. If all actors that have a relation to P are connected among each other, the value for aggregate constraint is high. Thus, individuals with high values for aggregate constraint leave structural holes if they refrain from engagement in a social network (Burt 1992, Burt 2004).[17] In particular, aggregate constraint is computed by acculmulating all dyadic constraint values for an actor $v_i$, which, in turn, are based on proportional strength. According to Burt (1992), proportional strength describes the importance of a link from one node to the other based on the number of connections a node has. In general, it is computed by dividing the sum of the weights of all connections between $v_i$ and $v_j$ by the sum of the weights of all connections $v_i$ has to other nodes.

$$\omega(v_i, v_j) = \frac{\omega_{ij} + \omega_{ji}}{\sum_k \omega_{ik} + \omega_{ki}} \qquad (5.1)$$

Since no further data for the weight of a connection was available, the weight of a link $\omega$ will always be one by definition. Figure 5.8 gives an example of the computation of proportional strength. Node A for instance, is directly connected to three other nodes. Proportional strength for each connection of A is therefore 0.33.

---

[17]For decades, structual holes have been of high interest to social network researchers because obtaining such holes is strategically beneficial as communication between other actors may be controlled and de novo entrants can achieve access to existing actors more easily (Burt 1992).

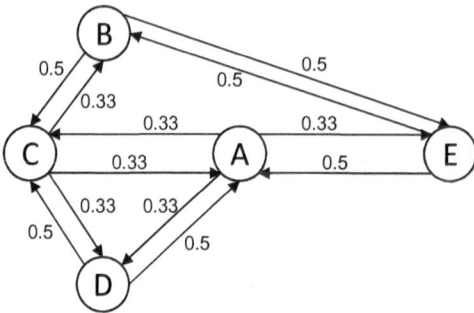

**Figure 5.8** Calculation of Proportional Strength

The calculation of proportional strength then is followed by the calculation of dyadic constraints. For a connection $ij$ between $v_i$ and $v_j$, each node $v_k$ is considered that complements a triad. For node A in Figure 5.8 the connection to D is not only realized directly, but also by taking into account the path through node C. Given equation 5.4, dyadic constraint for A to D is the sum of the direct and indirect weighted proportional strength, which then is $DC = (0.33 + (0.33 * 0.33))^2 = 0.1926$. Aggregated constraint is computed as the sum of all dyadic constraints for actor $v_i$ and usually exhibits values between 0 and 1. It therefore shows an actor's degree of involvement in the support links with alteris who are linked to each other by the same kind of bond (Di Nicola, Stanzani & Tronca 2011). If an actor has many contacts who are isolated from each other, the value of AGG-CONS is near 0, while if he has only one contact, it is 1 (Burt 1992).[18]

$$DC_{ij} = (PS_{ij} + \sum_{k \in I_i} PS_{ik} * PS_{kj})^2 \tag{5.2}$$

**BETW-CENTRALITY.** In accordance with Wassermann & Faust (1994), betweenness centrality takes into account that interactions between two nonadjacent actors might depend on the other actors in the set of actors. If an actor has connections to two actors who have no direct connection, the actor can mediate the communication between these two actors. Let $g_{jk}$ be the number of geodesics linking the two actors $j$ and $k$. If all paths are equally likely to be chosen, the likelihood of choosing a particular path is $1/g_{jk}$.

---

[18]For a more detailed description of aggregate constraint, see Burt (1992, p. 54ff), De Nooy et al. (2005, p. 133), Di Nicola et al. (2011, p. 8), Schneider (2009, p. 45), Stuckenschmidt & Klein (2004, p. 291), or Weinmann (2009, p. 36).

The distinct actor $i$ may be part of several communication paths between $j$ and $k$. The probability for $i$ to be part of a path then is $g_{jk}(n_i)/g_{jk}$. Betweenness centrality therefore is calculated as the sum of the estimated probabilities.

$$C_B(n_i) = \sum_{j<k} g_{jk}(n_i)/g_{jk} \tag{5.3}$$

In short, betweenness centrality measures the probability that the individual will fall on the shortest path between any two other individuals linked by email communication (Aral, Brynjolfsson & Van Alstyne 2007).

**PROX-PRESTIGE.** *Proximity prestige* is calculated by considering how afar actors are from the focal actor in average. All actors who can reach actor $v_i$ build what is commonly named the domain of $v_i$. The sum of all actors within the domain of $v_i$ is $I_i$. The average geodesic distance to $v_i$ than is calculated as:

$$D = \frac{\sum_{v_j \in I_i} d(v_j, v_i)}{|I_i|} \tag{5.4}$$

Thus, actors who cannot reach the focal actor through direct or indirect ways are not included. The value for proximity prestige is a decimal number between 0 and 1 and therefore ignores the number of ties an actor has, that is, an actor with 20 ties might achieve the same value as an actor with a single tie. In order to include the size of $v_i$'s influence domain $I_i$, $I_i$ is put in relation to the network size $n - 1$. Formula 5.2, thus represents the calculation for PROX-PRESTIGE in relation to the actor's network size.

$$P_P(v_i) = \frac{\frac{|I_i|}{n-1}}{\frac{\sum_{v_j \in I_i} d(v_j, v_i)}{|I_i|}} \tag{5.5}$$

**TECH-CONTRIBUTIONS** Most work on technical software communities uses CVS data[19] to operationalize technical contributions (e.g., Grewal et al. 2006). In contrast, I use the number of source files an individual has contributed to since authorship is also considered a proxy for technical contribution (Ghosh & David 2003, Tuomi 2004). In particular, this measure is calculated by considering any appearance of a person's name or e-mail address in the header of any source code file of the Linux kernel version 2.6.29.4.

---

[19]CVS stands for Concurrent Version System, a system that tracks developer contributions and uploads to a project (Shaikh & Cornford 2003).

Consequently, the variable can have values between 0 and the total number of source files with a .c ending in this kernel version, that is 11.443.

### 5.5.3.2 Independent Variables

In accordance with Dahlander & Wallin (2006), the independent variables are whether or not individuals have affiliations with firms that engage in the Linux kernel development. In order to separate individuals working for a firm from individuals who have no firm affiliation, I checked the e-mail postfix of their e-mail addresses. As outlined, individuals can send e-mails from multiple addresses. Thus, for the coding, I used the entries in the mapping table, that is, the single representation of multiple e-mail identities. As long as none of an individual's e-mail addresses was affiliated with a firm, I took the e-mail address an individual had sent most e-mails from. However, if one of different e-mail addresses for a person was a firm address, I took that firm address as his main address.

The coding of the variable SPONSORED is very detailed. In theory, a person with a firm e-mail address would be coded as 1 and a person without a firm e-mail address would be coded as 0. Unfortunately, reality is more complex and there are several cases where such a dichotomy is inappropriate. For example, firms that employ a business model dedicated to OSS might show different sponsoring behavior than firms with no OSS business model. Since Linux is an example of a community-initiated MVP, the single vendor business model is not applicable (Riehle 2009). However, although multiple firms engage in the development of Linux, they might pursue different strategies such as OSS as a complementary asset or an OSS business model (Grand et al. 2004). Therefore, I further checked if firms have a business model dedicated to OSS by justifying if the firm is listed as an official Linux supporter.

**Table 5.1** Coding of Variables

|  | Coding | Description | Example |
|---|---|---|---|
| null | no coding | not coded | |
| 0 | no firm affiliation | private, but unknown | @arndb.de |
| 1 | firm affiliation | firm, but no OSS BM | @addtoit.de |
| 2 | undefined | Kernel developers | @kernel.org |
| 3 | education | University, Research center | @harvard.edu |
| 4 | firm affiliation | firm with OSS BM | @intel.com |
| 5 | no firm affiliation | private, but known operator | @gmail.com |

Coding the entire database manually was impossible, so I used the following procedure. First, the 600 most active persons in terms of communication were identified. This is approximately the group of persons with 300 posts and more. These 600 individuals were categorized according to their e-mail postfix in the following way. Individuals with an e-mail address not affiliated with a firm and not affiliated with a big e-mail provider such as Yahoo, Gmail or GMX received the value 0. Category 1 embraces all individuals with a firm e-mail address that does not fall under category 4. 2 is reserved for individuals with e-mail addresses affiliated with Linux, such as "@kernel.org." All e-mail addresses that indicate educational institutions, such as "@mit.edu" or "@uni-koblenz.de", were put in category 3. Category 4 is reserved for individuals who have an e-mail address affiliated with a firm active in Linux development. I identified those firms that appear in the official Linux kernel report (Kroah-Hartmann et al. 2008) and that employ a business model dedicated to OSS. Finally, category 5 embraces all e-mail addresses affiliated with large e-mail providers. Based on the coding of the 600 most active individuals, the entire data set was then coded automatically. However, this implies that individuals with an e-mail address other than one of the 600 were not coded, meaning that they received the value *null*. Table 5.1 summarizes the coding scheme.

**Table 5.2** Aggregated Coding of Variables

| Name | 1 | 0 |
|---|---|---|
| NOT-SPONSORED | Not sponsored by any firm | otherwise |
| SPONS-NONDED | Sponsored by firm without OSS business model | otherwise |
| SPONS-DED | Sponsored by firm with OSS business model | otherwise |
| SPONSORED | Both sponsor types together | otherwise |

After running the first calculations, it turned out that this initial coding had to be adapted to various scenarios. For example, the group of individuals with e-mail addresses such as "@kernel.org" are responsible for a considerable number of technical contributions. However, persons who own such an e-mail address are affiliated with the Linux foundation, a non-profit organization. At first glance, these developers work for a non-profit organization and they seemingly might be volunteers. On the other hand, the Linux foundation is sponsored by firms, a fact that accounts for dependency on first interest. Thus, I put all individuals from category 2 into category 4.[20]

---

[20]I found support for this coding from representatives of RedHat at the 4th FLOSS Workshop in Jena

In order to test the hypotheses I transformed the coding into a dummy scheme. More precisely, I created variables that may take on values 0 and 1. For example, NOT-SPONS takes the value 1 if an individual is grouped in category 0, 2, or 5 and 0 otherwise. SPONS-NONDED refers to category 1, that is, firms with a business model not dedicated to OSS according to the Linux report. SPONS-DED instead refers to firms that are listed as active developers (category 4) and individuals working for the Linux foundation (category 2). Finally, SPONSORED combines both types of sponsorship in a single variable (Table 5.2).

### 5.5.3.3   Control Variables

Control variables are used to test if an effect is the sole consequence of the independent variable or not (Fahrmeir, Kneib & Lang 2007). Regarding the hypotheses, factors other than being sponsored by a firm may influence communication behavior and technical contribution. For example, individuals who are active in various domains ought to be more knowledgeable than their peers. They might know a solution to a problem as they were faced with the problem before in another context. In a similar vein, Miller, Fern & Cardinal (2007) have shown that the use of interdivisional knowledge positively affects the impact of an invention on subsequent technological developments. Individuals who are capable of combining knowledge from different domains are said to be boundary spanners (Tushman & Scanlan 1981a, Tushman & Scanlan 1981b). Within the Linux kernel data set, an individual is considered to be a boundary spanner if he is active in different mailing lists. BSPANNERS therefore refers to the number of different mailing lists an individual is active in.

In addition, a person's experience or tenure might put him in a position to be more active in communication. EXPERIENCE was therefore included as a control variable, measured as the time in days between the first and the last post in the mailing list. Maximum value is 3285 (9×365 days) since the data set was constructed from mailing list entries between 2000 and 2008.

The centrality measures I used do count for the number of communications an individual initiated or responded to. They do not take into account how many posts an individual actually made. Therefore, OVERALL-POSTS is the sum of all posts an individual ever sent to the mailing list and may be considered a measure for communication intensity.

---

since they confirmed that developers often use e-mail addresses such as "@kernel.org" in order to hide their real firm affiliation.

Considering technical contributions, I included a measure that counts for the number of co-authors an individual has (COAUTHORS). There is no example in the extant literature that suggests the use of such a measure. However, if an individual has a high number of co-authors, this might indicate that the amount of his actual contributions decreases. More precisely, if a person is not the sole author of a specific file, the work that is needed to produce the file is shared among different persons. As a consequence, the author has more time to dedicate himself to other files, which might increase his productivity measured in technical contributions.

Finally, although not of central interest, the standard deviation from Central European Time (CET) was calculated with regard to regional specifics (TIME-SD). However, I refused to include this measure into the analyses due to missing profound cause-effect relationships. While means, standard deviations, and correlations were calculated with the initial values, as suggested by Hahn et al. (2008), to avoid skewness, I used the natural log of BSPANNING and EXPERIENCE for further calculations.

## 5.6  Results

### 5.6.1  Part I: Sponsorship and Lateral Authority

As in Chapter 4, I used SPSS 18 to perform statistical calculations. Table 5.3 and Table 5.4 provide the descriptive statistics, such as means, standard deviations, and correlation coefficients. Not uncommon for high numbers of observations, the tables report high average correlation. About 10% of the individuals were identified as being sponsored either by a firm with a business model dedicated to Linux or by firms without a business model dependent on Linux.

Approximately a third of the individuals do not feature any firm affiliation.[21] This result mirrors the ongoing commercialization and increased firm engagement in OSS because comparable studies report smaller percentages. For example, based on a similar coding, Dahlander & Wallin (2006) report a mean for SPONS-DED of 0.023.[22] It is also worth reporting that the DEG-CENT (OUTDEGREE) and DEG-PRES (INDEGREE), two commonly used measures of actor centrality, both share a mean of 8.59 nodes and are highly correlated. Dahlander & Wallin (2006) provide identical values for both measures,

---

[21]Although dichotomous in nature, being sponsored or not does not sum up to 1 since not the entire dataset was coded.

[22]Dahlander & Wallin (2006) investigated GNOME, a Unix-like desktop environment.

too. OVERALL-POSTS exhibits values between 1 and 32727, and experience ranges from 0 to 3285 days.

Table 5.5 as well as Figure 5.9 provide a comparison of mean values between the four groups: SPONS-DED, SPONS-NONDED, NOT-SPONS, and UNCODED. As suggested, firm-sponsored individuals – both those sponsored by a firm with a business model towards Linux and those without – exhibit significantly higher values for TECH-CONTR, DEG-CENT, DEG-PRES, and OVERALL-POST, supported by a Duncan-Test as Post-Hoc-Test (Greene 2000).

Thus, individuals sponsored by a firm are generally more active in terms of communication and technical contribution. Concerning the difference between individuals sponsored by a firm with a business model dependent on Linux (SPONS-DED) and those with a business model not dependent on Linux (SPONS-NONDED), it is quite astonishing that those with a non-Linux business model do seek to form more ties and communicate more intensely (see Figure 5.9 upper right, lower left, and lower right). Individuals sponsored by a firm with a business model dependent on Linux only feature higher values in terms of technical contributions (upper left).

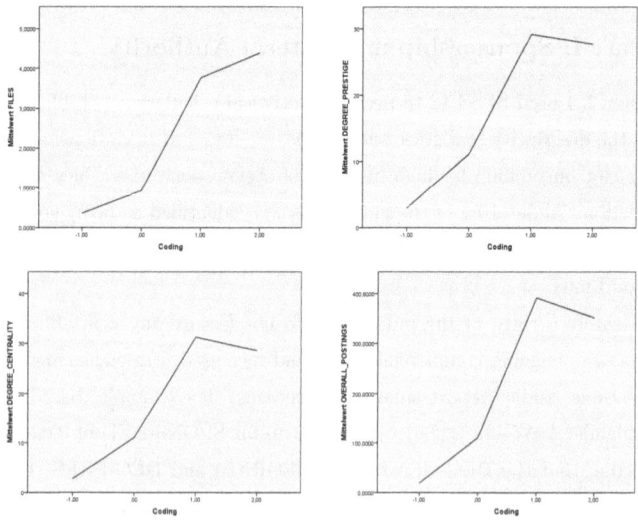

**Figure 5.9** Mean Comparison

Table **5.3** Linux Kernel 2006-2008: Descriptive Statistics and Correlations

| | Variables | N | Mean | S.D. | Min | Max | 1 | 2 | 3 | 4 | 5 |
|---|---|---|---|---|---|---|---|---|---|---|---|
| 1 | BSPANNING | 11899 | 2.741 | 5.163 | 1 | 111 | | | | | |
| 2 | COAUTHORS | 11899 | .276 | 1.212 | 0 | 27 | .278** | | | | |
| 3 | EXPERIENCE | 11899 | 625.779 | 973.061 | 0 | 3285 | .477** | .300** | | | |
| 4 | TECH-CONTRIBUTION | 11899 | .995 | 7.982 | 0 | 417 | .510** | .191** | .229** | | |
| 5 | OVERALL-POSTINGS | 11899 | 89.306 | 703.930 | 1 | 32727 | .785** | .188** | .250** | .532** | |
| 6 | TIME-SD | 11899 | 154.371 | 175.222 | 0 | 705 | .336** | .205** | .606** | .135** | .156** |
| 7 | AGG-CONSTRAINT | 11899 | .621 | .363 | 0 | 1.27 | -.389** | -.194** | -.446** | -.153** | -.180** |
| 8 | BETW-CENTRALITY | 11899 | .000 | .001 | 0 | .05 | .654** | .124** | .156** | .390** | .882** |
| 9 | DEG-CENTRALITY | 11899 | 8.590 | 43.210 | 0 | 1721 | .862** | .200** | .301** | .462** | .890** |
| 10 | DEG-PRESTIGE | 11899 | 8.590 | 40.375 | 0 | 1605 | .869** | .203** | .305** | .482** | .898** |
| 11 | EIG-CENTRALITY | 11899 | .001 | .009 | 0 | .43 | .703** | .157** | .205** | .432** | .901** |
| 12 | PRESTIGE | 11899 | .000 | 7.644 | -337 | 202 | -.282** | -.059** | -.089** | -.065** | -.284** |
| 13 | PROXIMITY-PRESTIGE | 11899 | .14 | .093 | 0 | 305 | .323** | .163** | .409** | .136** | .167** |
| 14 | SPONSORED | 11899 | .106 | .308 | 0 | 1 | .206** | .161** | .178** | .127** | .129** |
| 15 | NOT-SPONSORED | 11899 | .333 | .471 | 0 | 1 | .036** | -.017 | -.041** | -.006 | .021* |

**Table 5.4** Linux Kernel 2006-2008: Descriptive Statistics and Correlations (con'd)

| Variables | 6 | 7 | 8 | 9 | 10 | 11 | 12 | 13 | 14 |
|---|---|---|---|---|---|---|---|---|---|
| 6　TIME-SD | | | | | | | | | |
| 7　AGG-CONSTRAINT | -.556** | | | | | | | | |
| 8　BETW-CENTRALITY | .100** | -.123** | | | | | | | |
| 9　DEG-CENTRALITY | .218** | -.278** | .883** | | | | | | |
| 10　DEG-PRESTIGE | .225** | -.292** | .880** | .986** | | | | | |
| 11　EIG-CENTRALITY | .139** | -.170** | .827** | .872** | .886** | | | | |
| 12　PRESTIGE | -.043** | .031** | -.334** | -.447** | -.289** | -.249** | | | |
| 13　PROXIMITY-PRESTIGE | .559** | -.638** | .118** | .244** | .262** | .162** | .003** | | |
| 14　SPONSORED | .182** | -.215** | .085** | .158** | .161** | .124** | -.045** | .206** | |
| 15　NOT-SPONSORED | .032** | -.059** | .020* | .041** | .045** | .023* | .007 | .094** | -.244** |

$**p < .01; *p < .05$

Table **5.5** Comparing Means of Variables

| Variable | UNCODED | NOT-SPONS | SPONS-NONDED | SPONS-DED | F-Test |
|---|---|---|---|---|---|
| TECH-CONTR. | .36 | .93 | 3.73 | 4.38 | 90.98(.000) |
| DEG-CENT | 2.72 | 11.07 | 31.25 | 28.57 | 156.88(.000) |
| DEG-PRES | 2.93 | 11.14 | 29.17 | 27.72 | 162.31(.000) |
| OVERALL-POST. | 18.84 | 110.54 | 391.62 | 351.66 | 95.11(.000) |
| N | 6553 | 3967 | 393 | 986 | 11898 |

Tables 5.6.-5.9 report the hypothesis tests used ordinary least square (OLS) hierarchical regression analysis with centrality measures as dependent variables (Kuk 2006). Tables 5.10 and 5.11 provide results pertaining to technical contribution as the dependent variable. Hierarchical regression analysis is performed as a stepwise OLS regression analysis, where with each step more explanatory variables are included (Hair et al. 1998). Hierarchical regression is also commonly used to test for curvilinear models and U-shape relations (Tabachnik & Fidell 1989). In a good model, the explanatory power of the model, usually measured in $R^2$, should rise with any independent variable added (Lichtenthaler & Ernst 2009b).

All dependent variables in part I are measures of centrality suggesting issues of multicollinearity. However, multicollinearity is a consequence of high correlations among independent variables and leads to imprecise results only if it occurs within a single model. Thus, to test for multicollinearity between independent variables, variance inflation factor (VIF) was computed for each model (Hair et al. 1998). The values are all within an acceptable range not exceeding 1.675 (in case of natural log for BSPANNING) for all models 1A to 7D. The lowest VIF value was observed for NOT-SPONS with 1.079. Although regression analysis is said to be robust against violation of normal distribution prerequisites (Albers & Skiera 1999), all variables have been tested for normal distribution using the Kolmogoroff-Smirnoff-Test (Fasano & Franceschini 1987). The results suggest no significant deviation from the null hypotheses of variables being normally distributed, except for the centrality measures. All models were calculated with a random subsample of 10%, 15%, and 20% of the entire sample but revealed almost similar results.

Concerning the effect of sponsorship on network centrality, for each model, first only the control variables were tested (model type A). Then, in a stepwise procedure, the variables of interest were included with HOBBYISTS (model type B), followed by SPONS-NONDED (model type C) and SPONS-DED (model type D). As shown in models 1C and 2C in table 5.6, after controlling for the influence of BSPANNING and EXPERIENCE on DEG-CENT and DEG-PRES, the addition of SPONS-NONDED increases model 1C's and 2C's, respectively, explanatory power over model 1A and 2A ($\Delta R^2 = 0.001$). However, the effect is not strong. Thus, although including the variable SPONS-DED in models 1D and 2D shows that being sponsored by a firm with a business model dependent on Linux significantly explains DEG-CENT ($\beta = 0.021$) as well as DEG-PRES ($\beta = 0.023$), effects of other independent and control variables outperform SPONS-DED's influence.[23] H1 and H2 are therefore only partially supported.

Regarding H3 and H4, namely the effect of being sponsored on PRESTIGE and EIG-CENT, the results obtained provide partial support for H3 and support for H4. In particular, H3 predicts firm sponsored individuals to be less prestigious in the community. Although this does not hold true for individuals sponsored by a firm with a business model dependent on Linux (SPONSDED), a significant and negative effect is observable for individuals sponsored by a firm without a dedicated OSS business model (SPONS-NONDED), in partial support of H3. However, the explanatory power of model 3D is very weak with only about 3%. Supporting H4, the results show a significant effect of SPONS-DED on EIG-CENT.

H5 predicts individuals sponsored by a firm with a business model dependent on OSS will leave structural holes once they retire from community engagement. Crucial to this concept is the difference between being connected to well-connected ties and being connected to isolates. For example, if an individual is connected only to a single tie, he would receive a value for AGG-CONS of 1. Lower scores for AGG-CONS indicate greater potential for brokerage (De Nooy et al. 2005, Edwards & Crossley 2009). Thus, although model 6D only partially supports H6, namely the likelihood of seeking to obtain brokerage positions, model 5D fully supports H5 since being sponsored by a firm decreases the likelihood for high AGG-CONS values. In addition, model 5D exhibits high explanatory power with 41.7% of variance explained and features a difference $R^2$ of 0.006 for 5D over 5A, which is still low but the highest in the sample.

Finally, the results provided in Table 5.9 lead to a rejection of H7. In particular, they do not deliver any significant relation between the independent variable (SPONS-DED) and the dependent variable (PROX-PRES). The relation between SPONS-NONDED and PROX-PRES is significant, but was supposed to be positively correlated.

---

[23]The limited increase in explanatory power suggests a dominance of the control variables

**Table 5.6** Hierarchical OLS Regression I – DEGREE-CENTRALITY and DEGREE-PRESTIGE

| | Model 1A DEG-CENT | Model 1B DEG-CENT | Model 1C DEG-CENT | Model 1D DEG-CENT |
|---|---|---|---|---|
| BSPANNER | .624(.551)*** | .624(.551)*** | .619(.563)*** | .614(.563)*** |
| EXPERIENCE | -.159(.128)*** | -.159(.128)*** | -.159(.128)*** | -.161(.128)*** |
| NOT-SPONS | | .035(.708)*** | .040(.798)*** | .045(.732) |
| SPONS-NONDED | | | .037(.568)*** | .040(1.912)*** |
| SPONS-DED | | | | .021(1.300)** |
| Number of observations | 11898 | 11898 | 11898 | 11898 |
| $R^2$ adjusted | .290 | .291 | .292 | .293 |
| $\Delta R^2$ | | .001 | .002 | .003 |

| | Model 2A DEG-PRES | Model 2B DEG-PRES | Model 2C DEG-PRES | Model 2D DEG-PRES |
|---|---|---|---|---|
| BSPANNER | .627(.513)*** | .627(.512)*** | .622(.524)*** | .617(.524)*** |
| EXPERIENCE | -.153(.119)*** | -.153(.119)*** | -.153(.119)*** | -.155(.119)*** |
| NOT-SPONS | | .040(.685)*** | .044(.664)* | .049(.683)*** |
| SPONS-NONDED | | | .034(.729)*** | .037(1.779)*** |
| SPONS-DED | | | | .023(1.209)** |
| Number of observations | 11898 | 11898 | 11898 | 11898 |
| $R^2$ adjusted | .296 | .297 | .298 | .299 |
| $\Delta R^2$ | | .001 | .002 | .003 |

Note: ***$p < .001$; **$p < .01$; *$p < .05$; +$p < .1$. Standard errors in parentheses

Table 5.7 Hierarchical OLS Regression II – PRESTIGE and EIGENVECTOR - CENTRALITY

| | Model 3A PRESTIGE | Model 3B PRESTIGE | Model 3C PRESTIGE | Model 3D PRESTIGE |
|---|---|---|---|---|
| BSPANNER | −.217(.114)*** | −.217(.114)*** | −.213(.114)*** | −.214(.117)*** |
| EXPERIENCE | .091(.026)*** | .091(.026)*** | .091(.026)*** | .091(.027)*** |
| NOT-SPONS | | .009(.146) | .005(.148) | .006(.152) |
| SPONS-NONDED | | | −.030(.393)** | −.030(.396)** |
| SPONS-DED | | | | .004(.269) |
| Number of observations | 11898 | 11898 | 11898 | 11898 |
| R² adjusted | .030 | .030 | .031 | .031 |
| ΔR² | | .000 | .001 | .001 |

| | Model 4A EIG-CENT | Model 4B EIG-CENT | Model 4C EIG-CENT | Model 4D EIG-CENT |
|---|---|---|---|---|
| BSPANNER | .460(.000)*** | .460(.000)*** | .457(.000)*** | .456(.000)*** |
| EXPERIENCE | −.143(.000)*** | −.143(.000)*** | −.143(.000)*** | −.145(.000)*** |
| NOT-SPONS | | .019(.000)* | .022(.000)* | .029(.000)** |
| SPONS-NONDED | | | .024(.000)** | .028(.000)** |
| SPONS-DED | | | | .030(.000)** |
| Number of observations | 11898 | 11898 | 11898 | 11898 |
| R² adjusted | .150 | .150 | .150 | .151 |
| ΔR² | | .000 | .000 | .001 |

Note: ***$p < .001$; **$p < .01$; *$p < .05$; +$p < .1$, Standard errors in parentheses

**Table 5.8** Hierarchical OLS Regression III – AGG-CONSTRAINT and BETWEENNESS-CENTRALITY

| | Model 5A AGG-CONS | Model 5B AGG-CONS | Model 5C AGG-CONS | Model 5D AGG-CONS |
|---|---|---|---|---|
| BSPANNER | -.368(.004)*** | -.367(.004)*** | -.365(.004)*** | -.352(.004)*** |
| EXPERIENCE | -.342(.001)*** | -.343(.001)*** | -.343(.001)*** | -.338(.001)*** |
| NOT-SPONS | | -.058(.005)*** | -.060(.005)*** | -.073(.006)*** |
| SPONS-NONDED | | | -.014(.015)+ | -.020(.015)** |
| SPONS-DED | | | | -.056(.010)*** |
| Number of observations | 11898 | 11898 | 11898 | 11898 |
| R² adjusted | .411 | .414 | .414 | .417 |
| ΔR² | | .003 | .003 | .006 |

| | Model 6A BET-CENT | Model 6B BET-CENT | Model 6C BET-CENT | Model 6D BET-CENT |
|---|---|---|---|---|
| BSPANNER | .372(.000)*** | .372(.000)*** | .369(.000)*** | .369(.000)*** |
| EXPERIENCE | -.130(.000)*** | -.130(.000)*** | -.130(.000)*** | -.130(.000)*** |
| NOT-SPONS | | .017(.000)+ | .020(.000)* | .020(.000)* |
| SPONS-NONDED | | | .026(.000)** | .026(.000)** |
| SPONS-DED | | | | -.001(.000) |
| Number of observations | 11898 | 11898 | 11898 | 11898 |
| R² adjusted | .094 | .095 | .095 | .095 |
| ΔR² | | .001 | .001 | .001 |

Note: ***$p < .001$; **$p < .01$; *$p < .05$; +$p < .1$, Standard errors in parentheses

**Table 5.9** Hierarchical OLS Regression IV – PROX-PRESTIGE

|  | Model 7A PROX-PRES | Model 7B PROX-PRES | Model 7C PROX-PRES | Model 7D PROX-PRES |
|---|---|---|---|---|
| BSPANNER | $-.217(.114)^{***}$ | $-.217(.114)^{***}$ | $-.213(.114)^{***}$ | $-.214(.117)^{***}$ |
| EXPERIENCE | $.091(.026)^{***}$ | $.091(.026)^{***}$ | $.091(.026)^{***}$ | $.091(.027)^{***}$ |
| NOT-SPONS |  | $.009(.146)$ | $.005(.148)$ | $.006(.152)$ |
| SPONS-NONDED |  |  | $-.030(.393)^{**}$ | $-.030(.396)^{**}$ |
| SPONS-DED |  |  |  | $.004(.269)$ |
| Number of observations | 11898 | 11898 | 11898 | 11898 |
| $R^2$ adjusted | .030 | .031 | .031 | .032 |
| $\Delta R^2$ |  | .001 | .001 | .002 |

Note: $^{***}p < .001$; $^{**}p < .01$; $^{*}p < .05$; $^{+}p < .1$, Standard errors in parentheses

## 5.6.2    Part II: Sponsorship and Technical Contribution

Part II serves H8, H9, and H10. In each case, technical contribution is the dependent variable. Since different measures of centrality are used as independent variables, multicollinearity might be an issue. Computation of VIF leads to measures between 1.053 (SPONS-NONDED) and 1.805 (BSPANNING), except for the centrality measures DEG-CENT and DEG-PRES. With values of 34.959 and 35.260, respectively, both variables are above the recommended threshold of 10 that indeed suggests issues of multicollinearity (Hair et al. 1998). Running the regression without both variables, however, lead to comparable results, but the explanatory power decreases to values around 0.113.[24]

In Table 5.10, a significant increase in $R^2$ of 0.14 is observable from model 8A, which only includes the control variables and the full model 8D. As suggested, SPONS-DED is positive and significant while HOBBYISTS exhibits a negative impact on TECH-CONT. Thus, the findings well support H8 in that individuals sponsored by a firm with a business model dedicated to OSS are responsible for more technical contributions than their voluntary peers and their co-workers employed by firms without a business model dedicated to OSS.

Regarding H9 and H10, Table 5.11 provides results that support both hypotheses. In particular, communication intensity, measured as the total number of mails sent (OVERALL-POSTINGS) and DEG-PRES (i.e. INDEGREE in comparable studies) both affect the number of technical contributions positively in support of H10. Additionally, in line with H9, DEG-CENT (i.e. OUTDEGREE in comparable studies) has a negative influence on technical contributions. The difference in explanatory power between the full model and the control model is considerably high with 0.197. Table 5.12 provides a summary of all hypotheses for parts I and II.

---

[24]Mean centering may reduce multicollinearity, but is said to be effective in cases of mediating terms only (Aiken & West 1991, Box & Cox 1964, Carroll & Ruppert 1981, Echambadi & Hess 2007).

**Table 5.10** Hierarchical OLS Regression V – TECHNICAL CONTRIBUTION

| | Model 8A TECH-CONTR | Model 8B TECH-CONTR | Model 8C TECH-CONTR | Model 8D TECH-CONTR |
|---|---|---|---|---|
| BSPANNER | .327(.116)*** | .064(.123)*** | .064(.122)*** | .055(.124)*** |
| EXPERIENCE | −.074(.027)*** | −.010(.025) | −.010(.025) | −.013(.035) |
| COAUTHORS | .107(.060)*** | .086(.055)*** | .085(.055)*** | .082(.055)*** |
| DEG-CENTRALITY | | −.463(.009)*** | −.466(.009)*** | −.466(.009)*** |
| DEG-PRESTIGE | | .889(.009)*** | .894(.009)*** | .892(.009)*** |
| NOT-SPONS | | | −.026(.135)** | −.016(.140)[+] |
| SPONS-NONDED | | | | .015(.364)[+] |
| SPONS-DED | | | | .037(.248)*** |
| Number of observations | 11898 | 11898 | 11898 | 11898 |
| $R^2$ adjusted | .111 | .249 | .250 | .251 |
| $\Delta R^2$ | | .138 | .139 | .140 |

Note: ***$p < .001$; **$p < .01$; *$p < .05$; [+]$p < .1$, Standard errors in parentheses

**Table 5.11** Hierarchical OLS Regression VI – TECHNICAL CONTRIBUTION

| | Model 9A TECH-CONTR | VIF | Model 9B TECH-CONTR | VIF |
|---|---|---|---|---|
| BSPANNER | .327(.116)*** | 1.736 | .119(.119)*** | 2.326 |
| EXPERIENCE | −.074(.027)*** | 1.666 | −.006(.024) | 1.704 |
| COAUTHORS | .107(.060)*** | 1.112 | .070(.053)*** | 1.120 |
| OVERALL-POSTINGS | | | .564(.000)*** | 5.389 |
| DEG-CENTRALITY | | | −.549(.008)*** | 35.050 |
| DEG-PRESTIGE | | | .441(.009)*** | 38.623 |
| Number of observations | 11898 | | 11898 | |
| $R^2$ adjusted | .111 | | .308 | |
| $\Delta R^2$ | | | .197 | |

Note: ***$p < .001$; **$p < .01$; *$p < .05$; +$p < .1$, Standard errors in parentheses
VIF=Variance Inflation Factor

**Table 5.12** Summary of Hypotheses

| Hypothesis | Relation tested | supported |
|---|---|---|
| H1 | SPONSDED → DEG-CENT (OUTDEGREE) | partially |
| H2 | SPONSDED → DEG-PRES (INDEGREE) | partially |
| H3 | SPONSDED → PRESTIGE | partially |
| H4 | SPONSDED → EIG-CENT | yes |
| H5 | SPONSDED → AGG-CONS | yes |
| H6 | SPONSDED → BET-CENT | partially |
| H7 | SPONSDED → PROX-PRES | no |
| H8 | SPONSDED → TECH-CONTR | yes |
| H9 | DEG-CENT → TECH-CONTR | yes |
| H10 | DEG-PRES → TECH-CONTR | yes |

## 5.7    Discussion and Conclusion

In the preceding chapter, I argued that firms that intend to influence developments outside their boundaries might obtain RDBC or CBL. Whereas the investigation of Eclipse projects pertained to both forms of control, this chapter has been dedicated to CBL in particular. Leadership, however, may be achieved by different means such as being elected or being granted an explicit formal position. In the absence of hierarchical structures, such as in the case of OSS communities, firms can gain CBL by assigning their paid participants to achieve lateral authority by progressing to the center of the network (Dahlander & O'Mahony 2011). In this sense, this empirical study provides support for the concept of lateral authority and direct control in innovation communities external to the firm.

As the results show, individuals sponsored by firms are indeed more central than their voluntary peers, reflecting lateral authority. Of the ten hypotheses, five were fully supported, four were partially supported, and one hypothesis had to be rejected. However, the difference between individuals sponsored by a firm with a business model dedicated to OSS and those employed by firms without a dedicated OSS business model is not as present as expected. Building upon literature pertaining to control I suggested that firms with a business model dependent on OSS are most likely to secure their investments and to assign their employees to achieve positions of authority. In contrast, firms that support

OSS but do not have a dedicated OSS business have been predicted to be more central in a network than volunteers but less central than individuals sponsored by firms with a business model dependent on OSS. The majority of hypotheses did not support this suggestion. Most support derived from the analysis of sponsorship on aggregate constraint (Table 5.8) that supports the notion that individuals sponsored by a firm with a business model dependent on OSS are most likely to obtain brokerage positions. As De Nooy and his colleagues (2005, p. 134) put it:

> People or organizations with low aggregate constraint are hypothesized to perform better. It has been shown that employees with low constraint in organizations have more successful careers and that business sectors with lower constraint on firms are more profitable.

The findings on aggregate constraint therefore open various avenues for further research in the context of community structure, structural holes, career, and sponsoring.

In the hypotheses development section, I further argued that developers sponsored by a firm with a business model that depends on OSS are responsible for more technical contribution than those sponsored by firms with a business model not dependent on OSS. The results provide evidence for this suggestion. The full models in Table 5.10 provides a highly significant and positive effect ($\beta=0.037$) of being sponsored by a firm with a business model towards OSS and technical contribution. Moreover, not being sponsored by any firm significantly decreases the likelihood of being responsible for technical contributions. The model exhibits a value for $R^2$ of 0.251. Compared to the model with the control variables only (Model 8A), the $R^2$ value considerably increased ($\Delta R^2=0.140$).

Table 5.11 provides the results of the OLS regressions for the relation between network position and technical contribution. The full model (Model 9B) exhibits a $R^2$ value of 0.308, mirroring a good explanatory power. According to these results, seeking to form ties with other individuals decreases the probability of being responsible for technical contributions while being sought increases it.

Combining the findings for network centrality and technical contribution leads to reconciliation with prior work. Network theory posits that in the absence of hierarchical structures, network centrality is associated with power and authority (e.g., Bonacich 1987). Career experts further argue that lateral authority can be achieved by progressing to the center of the network (e.g., Berkowitz 1956). Consequently, individuals with an increased need for influencing the project, such as those sponsored by a firm, should be found in the center. However, individuals sponsored by a firm with a business model dedicated to OSS do not significantly differ from those sponsored by other firms in terms of their network position. Moreover, for some measures, individuals sponsored by a firm without a business model dependent on OSS even received higher values, that is, they are even more central (e.g., DEG-CENT, DEG-PRES).

Although the results concerning network positions were not as expected, they might be explained as follows. In the case of OSS, firms are generally unable to control individuals who are not embedded in their vertical control chains. They can only control participants whom they employ and whom they can confront with consequences in case of misbehavior. Lateral authority is most effective if an individual cannot pursue direct control or does not have access to formal leadership positions (Aghion & Tirole 1997, Leavitt 1951). However, as noted by Dahlander & O'Mahony (2011), firms do not need to control each individual within the community as long as they control the project. In the hypothesis development section, I argued that possibly the easiest way of controlling an OSSDP is obtaining leadership positions. Yet, contrary results show that contributing is the easiest way to control the project's trajectory. Individuals sponsored by firms with a business model dependent on OSS, though, do not need to gain lateral authority in the same manner as other sponsored individuals do, since they already possess the highest form of control: being allowed (and able) to contribute to the source code directly. Thus, firms with a business model not dependent on OSS or de novo entrants must engage in community work and communication more intensely – and thereby seek to gain positions of authority – because they are unable to control the project by their contributions.

In addition, the findings as such mirror the acceptance of firms within community-initiated OSSDPs. Whereas Dahlander & Wallin (2006) suggest that firms have to engage in communication work in order to gain legitimacy, the example of Linux shows that – five years after Dahlander & Wallin's (2006) investigation of GNOME – firms can afford to withdraw from initiating communication in favor of technical work. Admittedly, since firm-sponsored individuals are sought by knowledge seekers, they should not withdraw from answering communication requests. Model 9D provides support for this suggestion as being sought by peers is positively related to technical contribution.

To sum up, the data on Linux mailing lists and source code supports the distinction between lateral control and a more direct form of control by contributing. Both forms are appropriate means to control innovation projects located outside the boundaries of the firm. Concerning the management of OSS communities (cf. West & O'Mahony 2008), this study extends prior work by combining issues of communication work, technical contribution, and sponsorship what, to the best of my knowledge, has not yet been in the focus. In addition, the findings call for rethinking classical internal control chains for innovation beyond firm boundaries. Future research should therefore focus on the concept of dual careers and how employees can climb the ladder within the organizational boundaries while progressing to the center in an innovation community.

However, as with any research, the study features limitations that are worth noting. First, the data is taken entirely from the Linux kernel project (West & Dedrick 2001), which raises questions of generalizability. Future studies should therefore include different projects and compare the results with the findings provided here. Second, although a comparably elaborated coding of firm-sponsorship was used, e-mail data still possess certain disadvantages. In particular, a mapping table was used in order to separate firm-sponsored individuals from hobbyist and to control for multiple identities. However, it was only possible to code about 40% of the data set. Third, a causal relationship between being sponsored and network centrality was suggested, but firms can of course hire central individuals to work in alignment with firm interest. This calls for longitudinal research pertaining to changes in affiliation. Finally, the study did not take into account success measures, such as number of downloads or revenues of complementary products (Crowston, Howison & Annabi 2006, Ghanpanchi, Aurum & Low 2011, Lee, Kim & Gupta 2009). Thus, I call for future research that takes into account project success that best can be included in Structural Equation Model (SEM) (Andreev, Heart, Maoz & Pliskin 2009).

As such, the study provides a wealth of avenues for further investigations of open innovation communities (Armbruster 2005, Büchel & Armbruster 2006) in light of various theoretical considerations, such as organizational ambidexterity (e.g., Raisch et al. 2009), pecuniary and non-pecuniary absorptive and desorptive capacity (e.g., Lichtenthaler & Lichtenthaler 2009), modes of knowledge management and learning (e.g., Lee & Cole 2003), and network theory (e.g., Burt 1992).

# Chapter 6

# Summary, Conclusion, and Outlook

## 6.1 Theoretical and Empirical Contribution

The increasing prevalence of OSS as a format to manage innovation beyond firm boundaries with potentially reduced costs and increased access to external resources demands a better understanding of the options firms have to control what happens outside their walls. Thus, this dissertation started by asking:

> If innovation is managed at least partially outside and across the legal and organizational boundaries of the firm, how can a firm influence or even control a project its business model depends on without having discretionary power over developers external to the firm?

The pursuit of answers resulted in an extension of current management research to an area that has been somewhat neglected by information systems and management research, namely, the control of innovators external to the firm such as the developer community in the case of OSS. However, as shown in Chapter 2, the possibility of including external innovators into the innovation process only exist as a result of permeable firm boundaries. In this sense, this dissertation not only extends organizational control theory but contributes to the broader framework of open innovation and managing innovation beyond firm boundaries. In sum, the contribution of this dissertation pertains to three major aspects.

First, the discussion of relevant literature has led to an extension of one theoretical framework for knowledge management and one practical classification of OSS governance approaches. Regarding the knowledge management framework, I argued that the dimensions in the Lichtenthaler & Lichtenthaler (2009) framework, namely knowledge exploration, retention, and exploitation, might be complemented by a pecuniary dimension. More precisely, since research pertaining to knowledge management capacities such as absorptive and desorptive capacity (e.g., Cohen & Levinthal 1990, Lichtenthaler & Lichtenthaler 2010, Zahra & George 2002) does not differentiate a pecuniary from a non-pecuniary dimension, including a pecuniary perspective opens various avenues for further research (see Section 6.2).

In addition, a classification based on the differentiation between projects with one dominating firm (single-vendor projects; SVP) and those with more than one participating firm (multi-vendor projects; MVP)(Riehle 2011b, Schaarschmidt et al. 2011) as well as between projects initiated by a firm or community (Dahlander 2007), respectively, provides a systematization of governance modes that can be analyzed with regard to their control mechanisms. This categorization complements existing classifications basically in that it draws upon the distinction between firm-initiated SVP, community-initiated SVP, firm-initiated MVP, and community-initiated MVP. This distinction has been shown to be of importance not only to frame different OSS business models but to describe different governance modes, such as those known from transaction cost theory (i.e. Corporate Venture Capital, Joint Venture, Merger & Acquisition, etc.) (cf. Van de Vrande et al. 2009). The classification might help future investors to judge in which kind of OSS they invest and which risks come along with which type of OSS (cf. Chapter 4).

Second, based on the current debate about open innovation as a means to increase firms' performance, a theoretical gap concerning control has been identified. More precisely, from a transaction cost perspective, because firms increasingly deploy resources to OSSDPs they are likely to control the project's trajectory. However, it has been shown that organizational control theory is insufficient to explain and predict what firms do if developers are located beyond firms' vertical command chains. Since transactions between firms and voluntary developers are neither proceeded within hierarchies nor in the market, Demil & Lecocq (2006) coined the term bazaar governance to describe an alternative to the classical markets, hierarchies, and network governance forms (e.g., Jones, Hesterly & Borgatti 1997, Powell 1990). According to them, the peculiarity of bazaar governance is the simultaneous existence of low incentives intensity and low control intensity that in that form is considered new.

Other work on organizational control in OSS highlighted the difference between authority over collective work and authority over individuals, or simpler control over the project vs. control over the individual (Dahlander & O'Mahony 2011). In this sense, control intensity might only be low concerning individuals but might be high concerning the project itself. In a similar vein, Kuk (2006) emphasized that pure forms of bazaar governance are very seldom. Based upon this reasoning the concepts of "control by leadership" (CBL) and "resource-deployment-based control" (RDBC) as means to control innovation activities beyond firm boundaries have been developed. Capturing leadership positions within OSSDPs (CBL), a position that comes along with rights to contribute to the source code, as well as assigning their own developers to the OSSDP to work in line with firm interests (RDBC), thus are appropriate forms of control in that they internalize (parts of) the project.

Therefore, building on organizational control theory, which basically distinguishes formal from informal control mechanisms (Aghion & Tirole 1997, Aiken & Hage 1968, Kirsch 1997) as well as clan control from output and behavior control (Carver & Scheier 1981, Ouchi 1979, Rustagi et al. 2008, Snell 1992), this dissertation provides a better understanding of different control mechanisms firms can apply in different types of OSSDPs.

Finally, the third major contribution of this dissertation pertains to the empirical part. In both scenarios, the multi-project scenario (Eclipse) and the single-project scenario (Linux), I found support for the concepts of CBL and RDBC. For example, my analysis offers a strong support to the hypothesis that given limited positions of formal authority such as leadership, the more firms that engage within a project lead to an increased number of firm-sponsored developers, reflecting RDBC. As these developers have been socialized within the boundaries of the firm they bring the firm's norms and beliefs to the project and might pursue clan control (Ouchi 1979) beyond the boundaries of the firm.

Furthermore, contributing has been identified as possibly the highest form of controlling the OSSDP. In particular, although individuals without positions of formal authority might progress to the center of the community, a position that is said to be affiliated with power (e.g., Bonacich 1987), in order to control the project's trajectory, being able and allowed to contribute to the source code features the highest degree of control. However, the right to contribute is usually a result of prior technical and non-technical contribution to the project (e.g., Giuri et al. 2008).

## 6.2 Implications for Research

A prevalent theme in the strategic management and innovation literatures is that firms increasingly rely on outsiders both as a source of knowledge and as a means to commercialize it to gain and sustain competitive advantage (Boudreau & Lakhani 2009, Chesbrough 2003d, Teece 2000). However, absorbing as well as leveraging outsiders' knowledge demands specific knowledge management capacities (Lichtenthaler & Lichtenthaler 2009).

In the light of free external resources, I suggested to take into account a pecuniary perspective for each type of knowledge management activity, that is, knowledge exploration, retention, and exploitation (Lichtenthaler 2011c). Such a perspective raises interesting research questions. For example, from an economic perspective, firms should be likely to prefer access to free external resources over resources they have to pay for to obtain. However, the quality of resources as well as trust in these resources seem to be relevant moderators for acquisition decisions. Furthermore, the pecuniary perspective raises questions pertaining to knowledge flows. In particular, future research might investigate if and how freely absorbed knowledge differs from knowledge the firm has paid for regarding the knowledge retention and exploitation phase. I suggest knowledge protection in both the retention and exploitation phase to be lower for knowledge that has been absorbed without paying a monetary compensation as a consequence of the NIH-syndrome. Additionally, if the firm has paid for the knowledge or not should affect the firm's desorptive capacity at the organizational level (Lichtenthaler & Lichtenthaler 2010) as well as the sell-out attitude at the individual level (Lichtenthaler 2011c). These questions, however, can only be answered by further theoretical and empirical work.

A second avenue for further research stems from an observation that has been highlighted by Foss, Laursen & Pedersen (2011). They argue that despite a considerable number of research papers regarding user and open innovation, surprisingly little is said about the role of firms' internal organization for the success of including external knowledge into the innovation process. My notion of clan control beyond the boundaries of the firm addresses a similar aspect (cf. Kohli & Kettinger 2004, Ouchi 1979). In particular, if a firm assigns a relatively high number of developers to an OSSDP, these developers not only bring the firm's norms, beliefs, and values to the project but they also bring the project's norms, beliefs, and values to the firm (cf. Wallace 1995). In this sense, the firm adopts parts of the OSSDP's organization practices. Thus, further research should shed more light on the interaction between the firm's internal organization and the internal organization of the OSSDP it is dedicated to (cf. Henkel 2009, MacCormack et al. 2006) and how this relation affects both the firm's and the project's success.

Third, the results of this thesis have implications for research on division of labor and division of tasks (cf. Aiken & Hage 1968, Leijonhufvud 1986, Hauschildt & Chakrabarti

1989, Von Hippel 1990). Recent research on the structure of OSSDPs stressed that skill level and skill heterogeneity positively affect projects' survival and performances (Chou & He 2011, Giuri, Ploner, Rullani & Torrisi 2010). In a similar vein, various researchers emphasized that OSSDPs benefit from an internal division of labor between "specialists" and "generalists" as well as between "high-skilled" and "low-skilled" members (Giuri et al. 2010, Garzarelli & Fontanella 2011). Regarding firms and the commercialization of OSSDPs, Dahlander & Magnusson (2008) suggested that routine tasks could be outsourced to the community, while critical parts of software development could be pursued within the boundaries of the firm. My results complement this picture by providing evidence for the fact that in the case of MVPs, the community hardly exists of voluntary developers but of skilled personal of other firms. However, the minority of firms ought to be willing to assume the "community part" in that they only perform tasks that require low skills. Moreover, given equal commercial interests, firms are likely to obtain positions where they can execute tasks of complexity and importance that allow to influencing the project's trajectory. Therefore, I call for further research that investigates forms of division of labor between different firms in MVPs along with distributed leadership in innovation teams (Lindgren & Packendorff 2011).

Fourth, in line with aspects of division of labor between firms and the community, the results yield implications pertaining to the internal organization of an innovation community. Organization theorists argued that hierarchy has a curvilinear relation to effectiveness (e.g., Blau & Scott 1968, Krackhart 1993). In addition, Crowston & Howison (2006) showed that efficiency is closely related to the density of the developer network. If the network's density increases, efficiency decreases. This happens because individuals cannot increase their ability to respond to requests beyond a certain limit (in case the network density is a function of directed communication).

Relatedly, Kuk (2006) provided an analysis that showed that participation inequality is curvilinear related to knowledge sharing. Thus, neither a developer network where a few individuals are responsible for the majority of communication and work nor networks were communication activity is equally distributed performs well. Regarding my results, if many firms pursue RDBC and advice their own employees to actively contribute to discussions, participation inequality is likely to decrease. Theoretically, according to Kuk (2006), this should slow down knowledge sharing among participants and eventually the project's performance. Conversely, firm participation in OSSDPs is said to increase the project's performance (e.g., Chou & He 2011, Hahn et al. 2008). Berdou (2007, p. 199) consequently stressed the importance of considering how employment affects the division of labor in future studies:

> This challenges the idea that the social structure of F/OS communities is entirely dependent on internal project dynamics, since the ability to work on a project on a full-time basis inevitably affects the quality and quantity of individual contributions.

Future research therefore should compare different forms of internal OSSDP organization (e.g., hierarchical vs. flat, high degree of firm participation vs. low degree of firm participation) with regard to their performance.

Finally, this research has yielded implications to organizational control theory. By drawing attention to the fact that vertical command chains (that are usually supported by classical contractual control mechanism) are not effective in a scenario where volunteers dedicate their labor to an OSSDP, this study has provided support for two important concepts, namely resource-deployment-based control (RDBC) and control-by-leadership (CBL). Crucial for understanding both concepts is the fact that control over the project can be separated from control over individuals (Dahlander & O'Mahony 2011). In this sense, CBL pertains to control over the project while RDBC pertains to control over individuals. In the latter case, individuals are employed by the firm that intends to control the OSSDP. Thus, labor contract mechanisms are applicable within the boundaries of the firm. Within the community, developers who share the same employer then can pursue clan control, especially if those developers have yet been unable to capture leadership positions.

Thus, both RDBC and CBL revolve around binding developers, or the project, respectively, to the firm. Regarding the distinction of SVPs from MVPs, the single vendor searches for complementing free external resources (e.g., voluntary developers) that provide their service to the OSSDP in order to strengthen their development performance. Since firms that apply a single vendor approach usually own the entire rights to the source code, control over the project is not distributed among several actors. Consequently, those firms can act like traditional proprietary software vendors in terms of controlling the project and their employees. However, in order to encourage voluntary participation for a firm-driven SVP, the focal firm needs to communicate the benefit of participation for the volunteer or the community, respectively. Admittedly, such communication strategies can be rather costly (Dahlander & Magnusson 2005, O'Mahony 2003, West & O'Mahony 2008).

Conversely, in situations with multiple vendors that engage in the development of the OSSDP, several firms and volunteers constitute the community. The results of this study have shown that the degree of voluntary commitment in a MVP is a function of the project's history. Thus, in situations where the OSSDP was founded by firms, those firms that were active from the beginning as well as late entrants constitute what is considered a R&D consortia (e.g., Schaarschmidt & Von Kortzfleisch 2009) or an OSS service network (e.g., Feller, Finnegan, Fitzgerald & Hayes 2008). Since these projects aim at attracting participation of other firms, voluntary participation is very seldom.

Based upon these findings, several avenues for further research concerning control can be identified. First, from a theoretical point of view, future research should deliver additional support for the concepts of CBL and RDBC as means for controlling innovation activity which is pursued beyond firm boundaries. Second, future research should address the relation between CBL and RDBC more deeply. I suggested RDBC to be a form of control especially for de novo entrants that cannot appropriate from positions of authority yet, but evidence for this suggestion is still lacking. Third, future research might investigate how the suggested forms of control differ between MVPs with high degree and low degree of voluntary participation, a question that could not be entirely answered within this dissertation. Finally, future empirical research ought to develop new operationalizations of different forms of control beyond firm boundaries. In this study, the number of firm-sponsored developers and project leaders were used as measures for RDBC and CBL. Survey with reflective or formative constructs could yield additional insights (Petter, Straub & Rai 2007).

# 6.3    Implications for Management

As discussed in Section 3.3, firms use the OSS approach in various forms and for different reasons. Thus, managerial implications envisioned by the results of this dissertation are hardly generalizable and can pertain only to specific aspects in a given scenario. Furthermore, since this dissertation is subtitled "managing innovation beyond firm boundaries," the following managerial implications revolve around the firm as the level of analysis and do not address the individual, the community, or the foundation level. In sum, the managerial implications that derive ultimately from this study embrace three major areas, namely, tensions between firms and communities, firm control, and developers' education.

First, firms that engage in the development of OSS must accept that they tap into various tensions. On the one hand, if a software firm is able to encourage voluntary work, its development costs decrease because it does not have to pay for the external contributions.

On the other hand, since collaboration efficiency is a function of common interests, the benefits of improved innovative performance and reduced development costs will be outperformed by increased coordination and control costs if interests among different stakeholders become divergent. In other words, the OSSDP benefits from additional knowledge gains by opening innovation up, while diverse views on the project's trajectory leads to divergence (Almirall & Casadesus-Massanell 2010). Relatedly, West & O'Mahony (2008, p. 162) stated regarding these tensions:

> We showed that sponsors' community design decisions on these three dimensions reflected the inherent tension between two conflicting goals. On the one hand, firms wished to retain control over technologies fundamental to their business success. On the other hand, providing the opportunity structure for others to participate was a prerequisite for gaining the benefits from developing an external community.

Thus, firms that participate in such open innovation environments must know that these tensions cannot be dissolved, only balanced (Von Kortzfleisch 2004). For example, reducing the degree of divergence among actors by either excluding external participants or by increasing the number of actors with shared interests (e.g., firm-sponsored developers) will decrease access to external knowledge. In turn, fostering external contributions results in an increased level of divergence within the project. In this situation, firms have to give up parts of their control to enable the community to make decisions. Boudreau & Lakhani (2009) consequently highlighted the importance of the question if outside innovators should be organized as a collaborative community or as a competitive market. This question is very important for firms that intend to make use of the open source innovation approach (cf. Grand et al. 2004) because if the firm perceives a loss of control anyway, a market organization may be an alternative to complex community organization.

Second, firms should keep in mind that pursuing an OSS approach does not imply to get thousands of developers for free (Goldman & Gabriel 2005), who, moreover, directly align their work with the firm's interests. Extent research tells us that to get people to allocate their labor to a firm-driven OSSDP is challenging and requires a complex set of different incentives that meet different forms of motivation (e.g., Roberts et al. 2006, Shah 2006, Stewart et al. 2006). However, controlling those people once they have started to work for the OSSDP is even more challenging.

As discussed, it is possible to separate control over individuals from control over the project (Dahlander & O'Mahony 2011), a distinction that is mirrored by the introduced concepts of RDBC and CBL. By applying RDBC, firms achieve even a twofold benefit because it is possible to control developers by means of labor contracts within the boundaries of the firm and because those developers can bring the firm's norms and beliefs to the project. CBL, instead, aims at controlling the project's trajectory by capturing and occupying leadership positions rather than controlling individual developers. However, both forms of control share a downside that we already know from organizational control theory: Control is costly! In organizational settings, control costs derive from monitoring costs and costs of task definition. With the concepts of RDBC and CBL, additional costs arise. Whereas for RDBC, mirroring costs within the firm as well as training costs for pursuing clan control beyond the boundaries of the firm accrue, CBL is costly since designated firm-sponsored developers have to progress to the center of the developer network in order to achieve leadership positions, a procedure that might take years.

Thus, given that control is cost-intensive and control costs might exceed the benefits gained from OSS engagement, it is appropriate to ask: Do I really have to pursue control? Joel West recently posted in the open innovation blog after he listened to a speech of entrepreneur Eric McAfee and thereby provided an answer to the question:[1]

> For McAfee, the distinction between an entrepreneur and a manager is that an entrepreneur is someone "who allocates resources that they do not currently control," while the manager allocates resources they control. To me, this is the flip side of the oft-quoted Teece 1986 formulation. Teece focused on what entrepreneurs should do if they can't control resources. McAfee's point is that entrepreneurs often shouldn't even try — that it's usually better to buy or license the missing piece of the puzzle.
> [...]
> So if in Teece's world of 25 years ago, the goal was to control as many resources as possible and make do when you cannot, in McAfee's world, the goal is to control the resources that are important and partner for the rest. I think there are clearly cases when the latter approach is superior — particularly in a fast-moving industry where capital is scarce and the window of opportunity may close.

Finally, firms that intend to make use of the OSS development approach must invest in education and training of their personnel for various reasons. First, as other individuals that work in new product development, developers that work for software vendors are confronted with the NIH-syndrome (Katz & Allen 1982, Lichtenthaler & Ernst 2006). In this sense, assigning their own developers to work for an OSSDP, even if it was initiated

---

[1] "Open innovation, entrepreneurs, and resources", http://blog.openinnovation.net/2011/09/open-innovation-entrepreneurs-and.html, last access: 10/5/2011

by a firm, results in a twofold challenge: (1) Developers have to be prepared to align with the community's norms and beliefs in place and (2), they have to be prepared to introduce the firm's norms and beliefs into the community. This requires a mental change that only can be achieved by consecutive education and training. Additionally, managers must know that their employees pursue a dual career, one within the hierarchy of the employing firm and one within the community (Dahlander & O'Mahony 2011, Henkel 2006, Henkel 2009). Since dual careers might come along with dual identities (Wallace 1995), managers have to be careful with decisions concerning who of the internal development team they assign to work for an OSSDP (Daniel et al. 2011, Mehra, Dewan & Freimer 2011).

Second, qualifying developers for work in OSS communities is a necessity to prevent forks. A fork occurs if dissatisfied developers take the last version of the OSS product and start their own product line. Theoretically, this can happen at any time but mostly it is a direct consequence of community mismanagement. For example, ADempiere, an OSS Enterprise Resource Planning (ERP) system, was forked from the Compiere project due to disagreements between the developer community of Compiere and the for-profit firm ComPiere.Inc.[2]

Admittedly, a fork might not be the only threat of losing external development resources. Enlarged commercial activity within and around an OSSDP results in advanced productization (Feller et al. 2008, Somaya, Teece & Wakeman 2011), that is, stable release cycles, integration of customer requests, and service level agreements (offered by the firm, not by the community). However, as Berdou (2007, p. 198) noted: "the faster pace of development and shortened release cycles made it difficult for volunteers to absorb and keep up with changes and raised the barriers to participation for new developers." Thus, voluntary developers especially that do not work full time on a project might deallocate their resources from the project. While this might be intended in cases of MVPs where firms predominantly seek for commitments of other firms, this might be dangerous in an SVP scenario where voluntary committers are sought to complement own development resources.

In summary, this dissertation has shown that it is possible to influence and control OSSDPs even though parts of the development power are located beyond the firm's vertical command chains.

---

[2] ADempiere Homepage, http://adempiere.org/home/aboutus, last access: 11/2/2011

# Bibliography

Aamodt, A. & Nygard, M. (1995). Different roles and mutual dependencies of data, information, and knowledge: An AI perspective on their integration, *Data and Knowledge Engineering* **16**: 191–222.

Abdallah, F. & Wadhwa, A. (2009). Collaborating with our rivals: Identifying sources of coopetitive performance, *Proceedings of the DRUID Summer Conference*, Copenhagen, Denmark.

Abernathy, W. & Utterback, J. (1978). Patterns of industrial innovation, *Technology Review* **80**(7): 40–47.

Abrahamson, E. (1991). Managerial fads and fashions: The diffusion and rejection of innovations, *Academy of Management Review* **16**(3): 586–612.

Ågerfalk, P. & Fitzgerald, B. (2008). Outsourcing to an unknown workforce: Exploring opensourcing as a global sourcing strategy, *MIS Quarterly* **32**(2): 385–409.

Aghion, P. & Tirole, J. (1995). Some implications of growth for organizational form and ownership structure, *European Economic Review* **39**(3-4): 440–455.

Aghion, P. & Tirole, J. (1997). Formal and real authority in organizations, *Journal of Political Economy* **105**(1): 1–29.

Ahuja, G. & Katila, R. (2001). Technology acquisition and the innovation performance of acquiring firms: A longitudinal study, *Strategic Management Journal* **22**(3): 197–220.

Aiken, L. & West, S. (1991). *Multiple Regression: Testing and Interpreting Interactions*, Sage Publications, London, U.K.

Aiken, M. & Hage, J. (1968). Organizational interdependence and intra-organizational structure, *American Sociological Review* **33**: 631–652.

Albers, S. & Skiera, B. (1999). Regressionsanalyse, *in* A. Herrmann & C. Homburg (eds), *Marktforschung – Grundlagen, Methoden, Anwendungen*, Gabler, Wiesbaden, Germany, pp. 205–236.

Albert, S., Ashforth, B. & Dutton, J. (2000). Organizational identity and identification: Charting new waters and building new bridges, *Academy of Management Review* **25**: 13–17.

Alchian, A. & Demsetz, H. (1972). Production, information costs, and economic organization, *American Economic Review* **62**: 777–795.

Alexy, O. (2009). *Free Revealing: How Firms Can Profit From Being Open*, Gabler, Wiesbaden, Germany.

Allen, R. (1983). Collective invention, *Journal of Economic Behavior and Organization* **4**: 1–24.

Almirall, E. & Casadesus-Masanell, R. (2010). Open versus closed innovation: A model of discovery and divergence, *Academy of Management Review* **35**(1): 27–47.

Alston, L., Harris, E. & Mueller, B. (2009). De facto and de jure property rights: Land settlement and land conflict on the Australian, Brazilian and U.S. frontiers, *Australian National University, Discussion Papers Series No. 607*.

Alvesson, M. & Robertson, M. (2006). The best and the brightest: The construction, significance and effects of elite identities in consulting firms, *Organization* **13**: 195–224.

Amabile, T. (1985). Motivation and creativity: Effects of motivational orientation on creative writers, *Journal of Personality and Social Psychology* **48**: 393–399.

Amabile, T., Conti, R., Coon, H., Lazenby, J. & Herron, M. (1996). Assessing the work environment for creativity, *Academy of Management Journal* **39**(5): 1154–1184.

Amabile, T. & Gryskiewicz, N. (1989). The creative environment scales: Work environment inventory, *Creativity Research Journal* **2**: 231–253.

Amin, A. & Cohendet, P. (2000). Organisational learning and governance through embedded practices, *Journal of Management and Governance* **4**: 93–116.

Amin, A. & Cohendet, P. (2004). *Architectures of Knowledge: Firms, Capabilities and Communities*, Oxford University Press, Oxford, U.K.

Amit, R. & Schoemaker, P. (1993). Strategic assets and organizational rent, *Strategic Management Journal* **14**(1): 33–46.

Amit, R. & Zott, C. (2001). Value creation in e-business, *Strategic Management Journal* **22**(6/7): 493–520.

Andersen, B. & Konzelmann, S. (2008). In search of a useful theory of the productive potential of intellectual property rights, *Research Policy* **37**: 12–28.

Anderson, E. & Oliver, R. (1987). Perspectives on behavior-based versus output-based salesforce control systems, *Journal of Marketing* **51**(4): 76–88.

Andreev, P., Heart, T., Maoz, H. & Pliskin, N. (2009). Validating formative partial least square (PLS) models: Methodological review and empirical illustration, *Proceedings of the 30th International Conference on Information Systems (ICIS)*, Phoenix, AZ.

Arafat, O. & Riehle, D. (2009). The commit size distribution of open source software, *Proceedings of the 42nd Hawaiian International Conference on System Science (HICSS-42)*, Manoa, Hawaii.

Aral, S., Brynjolfsson, E. & Van Alstyne, M. (2007). Productivity effects of information diffusion networks, *Proceedings of the 28th International Conference on Information Systems (ICIS)*, Montreal, Canada.

Aral, S., Brynjolfsson, E. & Van Alstyne, M. (2012). Information, technology and information worker productivity, *Information Systems Research* **forthcoming**.

Armbruster, H. (2005). *Sozialstrukturen in Innovationsteams*, Gabler, Wiesbaden, Germany.

Arora, A. (1995). Licensing tacit knowledge: Intellectual property rights and the market for know-how, *Economics of Innovation and New Technology* **4**: 41–59.

Arora, A. & Gambardella, A. (2010). Ideas for rent: An overview of markets for technology, *Industry and Corporate Change* **19**(3): 775–803.

Arranz, N. & de Arroyabe, J. (2008). The choice of partners in R&D cooperation: An empirical analysis of Spanish firms, *Technovation* **28**: 88–100.

Arya, A., Glover, J. & Sivaramakrishnan, K. (1997). The interaction between decision control problems and the value of information, *Accounting Review* **72**(4): 561–574.

Audretsch, D. & Feldman, M. (1996). Knowledge spillovers and the geography of innovation and production, *The American Economic Review* **86**(3): 630–640.

Augustin, L. (2008). A new breed of P&L: The open source business financial model, Conference presentation given at "Open Source meets Business"-conference, January 23rd, 2008, Nuremberg, Germany.

Awazu, Y. & Desouza, K. (2004). Open knowledge management: Lessons from the open source revolution, *Journal of the American Society for Information Science and Technology* **55**(11): 1016–1019.

Bagozzi, R. & Dholakia, U. (2006). Open source software user communities: A study of participation in Linux user groups, *Management Science* **52**(7): 1099–1115.

Baldwin, C. & Clark, K. (2000). *Design Rules – The Power of Modularity*, MIT Press, Cambridge, MA.

Baldwin, C. & Clark, K. (2006). The architecture of participation: Does code architecture mitigate free riding in the open source development model?, *Management Science* **52**(7): 1116–1127.

Baldwin, C., Hienerth, C. & Von Hippel, E. (2006). How user innovations become commercial products: A theoretical investigation and case study, *Research Policy* **35**(9): 1291–1313.

Baldwin, C. & Von Hippel, E. (2009). Modeling a paradigm shift: From producer innovation to user and open collaborative innovation, *Harvard Business School Working Paper Series, No. 10-038* .

Barker, J. (1993). Tightening the iron cage: Concertive control in self-managing teams, *Administrative Science Quarterly* **38**(3): 321–341.

Barki, H. & Hartwick, J. (1994). Measuring user participation, user involvement, and user attitude, *MIS Quarterly* **18**(1): 59–82.

Barley, S. (1996). Technicians in the workplace: Ethnographic evidence for bringing work into organization studies, *Administrative Science Quarterly* **41**(3): 404–441.

Barley, S. & Kunda, G. (2001). Bringing work back in, *Organization Science* **12**: 76–95.

Barney, J. (1991). Firm resources and sustained competitive advantage, *Journal of Management* **17**(1): 99–120.

Barney, J. (2001). Resource-based *theories* of competitive advantage: A ten year retrospective on the resource-based view, *Journal of Management* **27**: 643–650.

Baron, S. (2010). Commentary: Statistics in marketing and consumer research, *Journal of Customer Behavior* **9**(3): 229–242.

Barthélemy, J. (2011). The Disney-Pixar relationship dynamics: Lessons for outsourcing vs. vertical integration, *Organizational Dynamics* **40**(1): 43–48.

Barzel, Y. (1989). *Economic Analysis of Property Rights*, Cambridge University Press, Cambridge, U.K.

Baskerville, R., Lyytinen, K., Sambamurthy, V. & Straub, D. (2010). A response to the design-oriented information systems research memorandum, *European Journal of Information Systems* .

Baskerville, R. & Myers, M. (2002). Information systems as a reference discipline, *MIS Quarterly* **26**(1): 1–14.

Büchel, B. & Armbruster, H. (2006). Erfolgsfaktoren von Innovationsteams: Der einfluss der übereinstimmenden Wahrnehmung zwischen Teammitgliedern und unternehmensinternen Stakeholdern, *Schmalenbachs Zeitschrift für betriebswirtschaftliche Forschung* **58**(6): 506–524.

Bechky, B. (2006). Gaffers, gofers, and grips: Role-based coordination in temporary organizations, *Organization Science* **17**(1): 3–21.

Behlendorf, B. (1999). Open source as a business strategy, *in* C. DiBona, S. Ockman & M. Stone (eds), *Open Sources: Voices from the Open Source Revolution*, O'Reilly and Associates, Cambridge, MA.

Bekkers, R., Duysters, G. & Verspagen, B. (2002). Intellectual property rights, strategic technology agreements and market structure: The case of GSM, *Research Policy* **31**: 1141–1161.

Belderbos, R., Faems, D., Leten, B. & Van Looy, B. (2010). Technological activities and their impact on the financial performance of the firm: Exploitation and exploration within and between firms, *Journal of Product Innovation Management* **27**(6): 869–882.

Bellman, R., Clark, C., Malcolm, D., Craft, C. & Ricciardi, F. (1957). On the construction of a multi-stage, multi-person business game, *Operations Research* **5**(4): 469–503.

Benbunan-Fich, R., Hiltz, S. & Turoff, M. (2002). A comparative content analysis of face-to-face vs. asynchronous group decision making, *Decision Support Systems* **34**(4): 457–469.

Berdou, E. (2007). *Managing the Bazaar: Commercialization and peripheral participation in mature commuity-led F/OS software projects*, PhD thesis, London School of Economics and Political Science, Department of Media and Communications, London, U.K.

Berdou, E. (2011). *Organization in Open Source Communities: At the Crossroads between the Gift and Market Economies*, Routledge, Oxford, U.K.

Bergquist, M. & Ljungberg, J. (2001). The power of gifts: Organizing social relationships in open source communities, *Information Systems Journal* **11**(4): 305–320.

Berkowitz, L. (1956). Personality and position, *Sociometry* **19**: 210–222.

Bessant, J. & Tidd, J. (2007). *Innovation and Entrepreneurship*, John Wiley & Sons, New York, NY.

Biggs, N. (1993). *Algebraic Graph Theory*, 2nd edn, Cambridge University Press, Cambridge, U.K.

Bird, A. (2003). Kuhn, nominalism, and empiricism, *Philosophy of Science* **70**: 690–719.

Bitzer, J. & Geishecker, I. (2010). Who contributes voluntarily to OSS? An investigation among German IT employees, *Research Policy* **39**: 165–172.

Bitzer, J., Schrettl, W. & Schröder, P. (2007). Intrinsic motivation in open source software development, *Journal of Comparative Economics* **35**: 160–169.

Blau, P. & Scott, W. (1962). *Formal Organizations*, Chandler Publishing, San Francisco, CA.

Boatright, J. (2004). Employee governance and ownership of the firm, *Business Ethics Quarterly* **14**(1): 1–21.

Bogers, M. (2011). The open innovation paradox: Knowledge sharing and protection in R&D collaborations, *European Journal of Innovation Management* **14**(1): 93–117.

Bolton, G., Ockenfels, A. & Ebeling, F. (2011). Information value and externalities in reputation building, *International Journal of Industrial Organization* **29**(1): 23–33.

Bonaccorsi, A., Giannangeli, S. & Rossi, C. (2006). Entry strategies under competing standards: Hybrid business models in the open source software industry, *Management Science* **52**(7): 1085–1098.

Bonaccorsi, A. & Rossi, C. (2006). Comparing motivations of individual programmers and firms to take part in the open source movement: From community to business, *Knowledge, Technology and Policy* **18**(4): 40–64.

Bonacich, P. (1972a). Factoring and weighting approaches to status scores and clique identification, *Journal of Mathematical Sociology* **2**: 113–120.

Bonacich, P. (1972b). A technique for analyzing overlapping memberships, *in* H. Costner (ed.), *Sociological Methodology*, Jossey-Boss, San Francisco, CA, pp. 176–185.

Bonacich, P. (1987). Power and centrality: A family of measures, *The American Journal of Sociology* **92**(5): 1170–1182.

Bonacich, P. (2007). Some unique properties of eigenvector centrality, *Social Networks* **29**: 555–564.

Borgatti, S. & Everett, M. (2006). A graph-theoretic perspective on centrality, *Social Networks* **28**: 466–484.

Borgatti, S., Everett, M. & Freeman, L. (2002). *UCInet for Windows: Software for Social Network Analysis*, Analytic Technologies, Harvard, MA.

Boudreau, K. (2008). Opening the platform vs. opening the complementary good? The effect on product innovation in handheld computing, *HEC working paper series, available at SSRN: http://ssrn.com/abstract=1251167* .

Boudreau, K. (2010). Open platform strategies and innovation: Granting access vs. devolving control, *Management Science* **56**(10): 1849–1872.

Boudreau, K. & Lakhani, K. (2009). How to manage outside innovation, *MIT Sloan Management Review* **50**(9): 69–76.

Bovet, D. & Cesati, M. (2002). *Understanding the Linux Kernel*, 2nd edn, O'Reilly, Sebastopol, CA.

Box, G. & Cox, D. (1964). An analysis of transformations, *Journal of the Royal Statistical Society* **Series B 26**(2): 211–252.

Brandenburger, A. & Nalebuff, B. (1996). *Coopetition*, Doubleday, New York, NY.

Brandenburger, A. & Stuart, H. (1996). Value-based business strategy, *Journal of Economics and Management Strategy* **5**: 5–25.

Brant, R. (1990). Assessing proportionality in the proportional odds model for ordinal logistic regression, *Biometrics* **46**: 1171–1178.

Brügge, B., Harhoff, D., Picot, A., Creighton, O., Fiedler, M. & Henkel, J. (2004). *Open Source Software. Eine ökonomische und technische Analyse*, Springer, Berlin, Germany.

Brockner, J., Spreitzer, G., Mishra, A., Hochwarter, W., Pepper, L. & Weinberg, J. (2004). Perceived control as an antidote to the negative effects of layoffs on survivor's organizational commitment and job performance, *Administrative Science Quarterly* **49**: 76–100.

Bromley, D. (1991). *Environment and Economy: Property Rights and Public Policy*, Blackwell Publishing, Cambridge, MA.

Brosius, F. (2006). *SPSS 14*, mitp, Heidelberg, Germany.

Brown, J. & Duguid, P. (2001). Organizational learning and communities-of-practice: Toward a unified view of working, learning and innovation, *Organization Science* **21**: 40–57.

Brusoni, S. & Prencipe, A. (2011). Patterns of modularization: The dynamics of product architecture in complex systems, *European Management Review* **8**(2): 67–80.

Brusoni, S., Prencipe, A. & Pavitt, K. (2001). Knowledge specialization, organizational coupling, and the firm: Why do firms know more than they make?, *Administrative Science Quarterly* **46**: 597–621.

Burrel, G. & Morgan, G. (1979). *Sociological Paradigms and Organisational Analysis: Elements of the Sociology of Corporate Life*, Heinemann, London, U.K.

Burt, R. (1992). *Structural Holes: The Social Structure of Competition*, Harvard University Press, Cambridge, MA.

Burt, R. (2004). Structural holes and good ideas, *American Journal of Sociology* **110**(2): 349–399.

Burt, R. (2007). Brokerage and closure: An introduction to social capital, *European Sociological Review* **23**(5): 666–667.

Buxmann, P., Diefenbach, H. & Hess, T. (2008). *Die Softwareindustrie – Ökonomische Prinzipien, Strategien, Perspektiven*, Springer, Berlin, Germany.

Capra, E., Francalanci, C., Merlo, F. & Rossi Lamastra, C. (2011). Firms' involvement in open source projects: A trade-off between software structural quality and popularity, *Journal of Systems and Software* **84**(1): 144–161.

Capra, E. & Wasserman, A. (2008). A framework for evaluating managerial styles in open source projects, *Proceedings of the 4th International Conference on Open Source Systems (OSS)*, Milan, Italy.

Cardinal, L. (2001). Technological innovation in the pharmaceutical industry: The use of organizational control in managing research and development, *Organization Science* **12**(1): 19–36.

Cardinal, L., Sitkin, S. & Long, C. (2004). Balancing and rebalancing in the creation and evolution in organizational control, *Organization Science* **15**(4): 411–431.

Carlile, P. (2002). A pragmatic view of knowledge and boundaries: Boundary objects in new product development, *Organization Science* **13**(4): 442–455.

Carlile, P. (2004). Transferring, translating, and transforming: An integrative framework for managing knowledge across boundaries, *Organization Science* **15**(5): 555–568.

Carroll, M. (2006). Creative commons and the new intermediaries, *Michigan State Law Review 2006* (1): 45–65.

Carroll, M. (2007). Creative commons as conversational copyright, *in* P. K. Yu (ed.), *Intellectual Property and Information Wealth. Issues and Practices in the Digital Age, Vol. 1*, Praeger Publishers, pp. 445–461.

Carroll, R. & Ruppert, D. (1981). On prediction and the power transformation family, *Biometrika* **68**: 609–615.

Carson, S., Madhok, A., Varman, R. & John, G. (2003). Information processing moderators of the effectiveness of trust-based governance in interfirm R&D collaboration, *Organization Science* **14**(1): 45–56.

Carver, C. & Scheier, M. (1981). *Attention and Self-Regulation: A Control Theory Approach to Human Behavior*, Springer, New York, NY.

Casadesus-Masanell, R. & Ghemawat, P. (2006). Dynamic mixed duopoly: A model motivated by linux vs. windows, *Managememt Science* **52**(7): 1072–1084.

Casadesus-Masanell, R. & Llanes, G. (2009). Mixed source, *Harvard Business School Working Paper Series, No. 10-022*.

Casadesus-Masanell, R. & Ricart, J. (2010). From strategy to business model and onto tactics, *Long Range Planning* **43**(2/3).

Cassiman, B. & Veugelers, R. (2006). In search of complementarity in innovation strategy: Internal R&D and external knowledge acquisition, *Management Science* **52**(1): 68–82.

Cecez-Kecmanovic, D. (2011). On methods, methodologies and how they matter, *Proceedings of the 19th European Conference on Information Systems (ECIS)*, Helsinki, Finland.

Chandler, A. (1962). *Strategy and Structure – Chapters in the History of Industrial Enterprise*, MIT Press, Cambridge, MA.

Chandler, A. (1992). Organizational capabilities and the eonomic history of the industrial enterprise, *Journal of Economic Perspectives* **6**: 79–100.

Chatain, O. & Zemsky, P. (2011). Value creation and value capture with frinctions, *Strategic Management Journal* **32**(11): 1206–1242.

Chatterjee, S. (2001). Informations systems research and relevance, *Communications of the AIS* **6**(8): 1–7.

Chen, M., Iyigun, M. & Maskus, K. (2007). General public licensing and the intensity of aggregate software development, *Economics of Innovation and New Technology* **16**(5-6): 451–466.

Chen, W. & Hirschheim, R. (2004). A paradigmatic and methodological examination of information systems research from 1991 to 2001, *Information Systems Journal* **14**(3): 197–235.

Chengalur-Smith, I., Nevo, S. & Demertzoglou, P. (2010). An empirical analysis of the business value of open source infrastructure technologies, *Journal of the Association of Information Systems* **11**(Special Issue): 708–729.

Chengalur-Smith, I., Sidorova, S. & Daniel, S. (2010). Sustainability of free/libre open source projects: A longitudinal study, *Journal of the Association of Information Systems* **11**(Special Issue): 657–683.

Chesbrough, H. (2002). Graceful exits and foregone opportunities: Xerox's management of its technology spinoff organizations, *Business History Review* **76**(4): 803–838.

Chesbrough, H. (2003a). The era of open innovation, *MIT Sloan Management Review* **44**(3): 35–41.

Chesbrough, H. (2003b). The governance and performance of Xerox's technology spin-off companies, *Research Policy* **32**: 403–421.

Chesbrough, H. (2003c). The logic of open innovation: Managing intellectual property, *California Management Review* **45**(3): 33–58.

Chesbrough, H. (2003d). *Open Innovation*, Harvard University Press, Cambridge, MA.

Chesbrough, H. (2006). *Open Business Models – How to Thrive in the New Innovation Landscape*, Harvard Business School Press, Cambridge, MA.

Chesbrough, H. (2010). Business model innovation: Opportunities and barriers, *Long Range Planning* **43**(2/3): 354–363.

Chesbrough, H. (2011). *Open Service Innovation: Rethinking Your Business to Grow and Compete in a New Area*, Jossey-Boss, San Francisco, CA.

Chesbrough, H. & Rosenbloom, R. (2002). The role of the business model in capturing value from innovation: Evidence from Xerox corporation's technology spin-off companies, *Industrial and Corporate Change* **11**(3): 529–555.

Chesbrough, H., Vanhaverbeke, W. & West, J. (2006). *Open Innovation: Researching a New Paradigm*, Oxford University Press, Oxford, U.K.

Chiaroni, D., Chiesa, V. & Frattini, F. (2009). Investigating the adoption of open innovation in the biopharmaceutical industry: A framework and an empirical analysis, *European Journal of Innovation Management* **12**(3): 285–305.

Chiaroni, D., Chiesa, V. & Frattini, F. (2010). Unraveling the process from closed to open innovation: Evidence from mature, asset-intensive industries, *R&D Management* **40**(3): 222–245.

Chidamber, S. & Kon, H. (1994). A research retrospective of innovation inception and success: The technology-push, demand-pull question, *International Journal of Technology Management* **9**(1): 94–112.

Chou, S.-W. & He, M.-Y. (2011). The factors that affect the performance of open souirce software development – The perspective of social capital and expertise integration, *Information Systems Journal* **21**: 195–219.

Choudhury, V. & Sabherwal, R. (2003). Portfolios of control in outsourced software development projects, *Information Systems Research* **14**(3): 291–314.

Christensen, C., Cook, S. & Hall, T. (2005). Marketing malpractice – the cause and the cure, *Harvard Business Review* **73**(12).

Coase, R. (1937). The nature of the firm, *Economica* **4**: 386–405.

Coase, R. (1960). The problem of social cost, *Journal of Law and Economics* **3**: 1–44.

Coase, R. (1988). *The Firm, the Market, and the Law*, University of Chicago Press, Chicago, IL.

Coff, R. (2003). Bidding wars over R&D-intensive firms: Knowledge, opportunism, and the market for corporate control, *Academy of Management Journal* **46**(1): 74–85.

Cohen, J. (1988). *Statistical Power Analysis for the Behavioral Sciences*, 2nd edn, Lawrence Erlbaum Associates, Hillsdale.

Cohen, W. & Levinthal, D. (1990). Absorptive capacity: A new perspective on learning and innovation, *Administrative Science Quarterly* **35**(1): 128–152.

Colombo, M., Rabbiosi, L. & Reichstein, T. (2010). Organizing for external knowledge sourcing, *European Management Review* **7**: 74–76.

Conger, J. & Kanungo, R. (1987). Toward a theory of charismatic leadership in organizational settings, *Academy of Management Review* **12**(4): 637–647.

Conway, M. (1968). How do committee's invent?, *Datamation* **14**(5): 28–31.

Cooper, R. (1994). Third-generation new product processes, *Journal of Product Innovation Management* **11**(1): 3–14.

Corbet, J., Kroah-Hartmann, G. & McPherson, A. (2010). Linux kernel development: How fast it is going, who is doing it, what they are doing, and who is sponsoring it, *Technical report*, Linux Foundation, San Francisco, CA.

Cornford, T., Shaikh, M. & Ciborra, C. (2010). Hierarchy, laboratory and collective: Unveiling Linux as innovation, machination and constitution, *Journal of the Association for Information Systems* **11**(12): 809–837.

Cropanzano, R. & Mitchell, M. (2005). Social exchange theory: An interdisciplinary review, *Journal of Management* **31**(6): 874–900.

Cross, R. & Prusak, L. (2002). People who make organizations go-and stop, *Harvard Business Review* **80**(6): 104–112.

Crowston, K. (1997). A coordination theory approach to organizational process design, *Organization Science* **8**(2): 157–175.

Crowston, K. & Howison, J. (2006). Hierarchy and centralization in free and open source software team communications, *Knowledge, Technology and Policy* **18**: 65–85.

Crowston, K., Howison, J. & Annabi, H. (2006). Information systems success in free and open source software development: Theory and measures, *Software Process: Improvement and Practice* **11**(2): 123–148.

Dafermos, G. (2001). Management and virtual decentralized networks: The Linux project, *First Monday* **6**(11).

Dahan, E. & Hauser, J. (2002). The virtual customer, *Journal of Product Innovation Mangement* **19**(5): 332–353.

Dahlander, L. (2005). Appropriation and appropriability in open source software, *International Journal of Innovation Management* **9**(3): 259–285.

Dahlander, L. & Frederiksen, L. (2011). The core and cosmopolitans: A relational view of innovation in user communities, *Organization Science* **forthcoming**.

Dahlander, L. & Gann, D. (2010). How open is innovation?, *Research Policy* **39**: 699–709.

Dahlander, L. & Magnusson, M. (2005). Relationships between open source software companies and communities: Observations from Nordic firms, *Research Policy* **34**(4): 481–493.

Dahlander, L. & Magnusson, M. (2008). How do firms make use of open source communities?, *Long Range Planning* **41**: 629–649.

Dahlander, L. & O'Mahony, S. (2011). Progressing to the center: Coordinating project work, *Organization Science* **22**(4): 961–979.

Dahlander, L. & Wallin, M. (2006). A man on the inside: Unlocking communities as complementary assets, *Research Policy* **35**: 1243–1259.

Dalton, M. (1959). *Men Who Manage: Fusions of Feeling and Theory in Administration*, Wiley, New York, NY.

Daniel, S., Maruping, L., Cataldo, M. & Herbsleb, J. (2011). When cultures clash: Participation in open source communities and its implications for organizational commitment, *Proceedings of the 31st International Conference on Information Systems (ICIS)*, Shanghai, China.

Danziger, K. (1979). The positivist repudiation of Wundt, *Journal of the History of the Behavioral Sciences* **15**: 205–230.

Darmon, E. & Torre, D. (2009). Open source and commercial software platforms: Is coexistence a temporary or sustainable outcome?, *International Journal of Open Source Software & Processes* **1**: 67–80.

Davenport, S. & Daellenbach, U. (2011). 'Belonging' to a virtual research centre: Exploring the influence of social capital formation processes on member identification in a virtual organization, *British Journal of Management* **22**(1): 54–76.

Davenport, T. & Prusak, L. (1998). *Working Knowledge: How Organizations Manage What They Know*, Harvard Business School Press, Cambridge, MA.

De Jong, J. & Von Hippel, E. (2009). Transfers of user process innovations to process equipment producers: A study of Dutch high-tech firms, *Research Policy* **38**: 1181–1191.

De Laat, P. (2005). Copyright or copyleft? An analysis of property regimes for software development, *Research Policy* **34**(10): 1511–1532.

De Laat, P. (2007). Governance of open source software: State of the art, *Journal of Management and Governance* **11**(2): 165–177.

De Nooy, W., Mrvar, A. & Batagelji, V. (2005). *Exploratory Social Network Analysis with Pajek*, Series: Structural Analysis in the Social Sciences, Cambridge University Press, Cambridge, U.K.

De Rond, M. & Bouchikhi, H. (2004). On the dialectics of strategic alliances, *Organization Science* **15**(1): 56–69.

De Vaujany, F.-X., Lesca, N., Fomin, V. & Loebbecke, C. (2008). The espoused theories of IS: A study of general editorial statements, *Proceedings of the 29th International Conference on Information Systems (ICIS)*, Paris, France.

Deci, E. (1971). Effects of externally mediated rewards on intrinsic motivation, *Journal of Personality and Social Psychology* **18**: 105–115.

Deci, E., Koestner, R. & Ryan, R. (1999). A meta-analytic review of experiments examining the effects of extrinsic rewards on intrinsic motivation, *Psychological Bulletin* **125**(6): 627–668.

Deci, E. & Ryan, R. (1980). The empirical exploration of intrinsic motivational processes, *in* L. Berkowitz (ed.), *Advances in Experimental Social Psychology Vol. 13*, Academic Press, New York, NY, pp. 39–80.

Deci, E. & Ryan, R. (1985). *Intrinsic Motivation and Self-Determination in Human Behavior*, Plenum Press, New York, NY.

Deci, E. & Ryan, R. (1987). The support of autonomy and the control of behavior, *Journal of Personality and Social Psychology* **53**: 1024–1037.

Deek, F. & McHugh, J. (2007). *Open Source: Technology and Policy*, Cambridge University Press, Cambridge, U.K.

Demil, B. & Lecocq, X. (2006). Neither market nor hierarchy nor network: The emergence of bazaar governance, *Organization Studies* **27**(10): 1447–1466.

Demsetz, H. (1967). Toward a theory of property rights, *The American Economic Review* **57**(2): 347–359.

Demsetz, H. (1988). *Ownership, Control, and the Firm. Vol. I*, Blackwell Publishing, Oxford, U.K.

Dewett, T. (2007). Linking intrinsic motivation, risk taking, and employee creativity in an R&D environment, *R&D Management* **37**(3): 197–208.

Di Nicola, P., Stanzani, S. & Tronca, L. (2011). Personal networks as social capital: A research strategy to measure contents and forms of social support, *Italian Sociological Review* **1**(1): 1–15.

Dierickx, I. & Cool, K. (1989). Asset stock accumulation and sustainability of competitive advantage, *Management Science* **35**(12): 1504–1511.

Doeringer, P. & Piore, M. (1971). *Internal Labor Markets and Manpower Analysis*, D.C. Heath and Company, Lexington, MA.

Doganova, L. & Eyquem-Renault, M. (2009). What do business models do? Innovation devices in technology entrepreneurship, *Research Policy* **38**(10): 1559–1570.

Dreier, T. & Nolte, G. (2006). Einführung in das Urheberrecht, *in* J. Hofmann (ed.), *Wissen und Eigentum. Geschichte, Recht und Ökonomie stoffloser Güter*, Bundeszentrale für politische Bildung, Bonn, Germany, pp. 53–54.

Druskat, V. & Wheeler, J. (2003). Managing from the boundary: The effective leadership of self-managing work teams, *Academy of Management Journal* **46**(4): 435–457.

Dushnitsky, G. & Klueter, T. (2011). Is there an eBay for ideas? Insights from online knowledge marketplaces, *European Management Review* **8**(1): 17–32.

Dushnitsky, G. & Lenox, M. (2006). When does corporate venture capital investment create firm value?, *Journal of Business Venturing* **21**(6): 753–772.

Easterby-Smith, M. & Prieto, I. (2008). Dynamic capabilities and knowledge management: An integrative role for learning?, *British Journal of Management* **19**(3): 235–249.

Echambadi, R. & Hess, J. (2007). Mean-centering does not alleviate collinearity problems in moderated multiple regression models, *Marketing Science* **26**: 438–445.

Economides, N. & Katsamakas, E. (2006). Two-sided competition of proprietary vs. open source technology platforms and implications for software industry, *Management Science* **52**(7): 1057–1071.

Edmondson, A. & McManus, S. (2007). Methodological fit in management field research, *Academy of Management Review* **32**(4): 1155–1179.

Edvardsson, B., Tronvoll, B. & Gruber, T. (2011). Expanding understanding of service exchange and value co-creation: A social construction approach, *Academy of Marketing Science* **39**: 327–339.

Edwards, G. & Crossley, N. (2009). Measures and meanings: Exploring the Ego-net of Helen Kirkpatrick Watts, militant suffragette, *Methodological Innovations Online* **4**: 37–61.

Eisenhardt, K. (1985). Control: Organizational and economics approaches, *Management Science* **31**(2): 134–149.

Eisenhardt, K. (1989). Agency theory. an assessment and review, *Academy of Management Review* **14**(1): 57–74.

Eisenmann, T., Parker, G. & Van Alstyne, M. (2008). Opening platforms: How, when and why?, *Harvard Business School Working Paper Series, No. 09-030* .

Elias, S. (2009). Restrictive versus promotive control and employee work outcomes: The moderating role of locus of control, *Journal of Management* **35**(2): 369–392.

Elmquist, M., Fredberg, T. & Ollilia, S. (2009). Exploring the field of open innovation, *European Journal of Innovation Management* **12**(3): 326–345.

Enkel, E., Gassmann, O. & Chesbrough, H. (2009). Open R&D and open innovation: Exploring the phenomenon, *R&D Management* **39**(4): 311–316.

Erwin, D. & Krakauer, D. (2004). Insights into innovation, *SCIENCE* **304**(5674): 1117–1119.

Ethiraj, S. & Levinthal, D. (2004). Modularity and innovation in complex systems, *Management Science* **50**(2): 159–173.

Fahrmeir, L., Kneib, T. & Lang, S. (2007). *Regression - Modelle. Methoden, Anwendungen*, Springer, Berlin, Germany.

Fang, C., Lee, J. & Schilling, M. (2010). Balancing exploration and exploitation through structural design: The isolation of subgroups and organizational learning, *Organization Science* **21**: 625–642.

Fasano, G. & Franceschini, A. (1987). A multidimensional version of the Kolmogorov-Smirnov test, *Royal Astronomical Society: Monthly Notices* **225**(3): 155–170.

Faust, K. (1997). Centrality in affiliation networks, *Social Networks* **19**: 249–282.

Feller, J., Finnegan, P., Fitzgerald, B. & Hayes, J. (2008). From peer production to productization: A study of socially enabled business exchange in open source service networks, *Information Systems Research* **19**(4): 475–493.

Fershtman, C. & Gandal, N. (2007). Open source software: Motivation and restrictive licensing, *Journal of International Economics and Economic Policy* **4**(2): 209–225.

Fershtman, C. & Gandal, N. (2011). Direct and indirect knowledge spillovers: The "social network" of open-source projects, *RAND Journal of Economics* **42**(1): 70–91.

Fey, C. & Birkinshaw, J. (2005). External sources of knowledge, governance mode and R&D performance, *Journal of Management* **31**(4): 597–621.

Fichman, R. (2000). The diffusion and assimilation of information technology innovation, *in* R. Zmud (ed.), *Framing the Domains of IT Management: Projecting the Future...Through the Past*, Pinnaflex Publishing, Cincinnati, OH.

Field, A. (2005). *Discovering Statistics Using SPSS*, 2nd edn, Sage Publication, London, U.K.

Fitzgerald, B. (2006). The transformation of open source software, *MIS Quarterly* **30**(3): 587–598.

Flatten, T., Engelen, A., Zahra, S. & Brettel, M. (2011). A measure for absorptive capacity: Scale development and validation, *European Management Journal* **29**(2): 98–116.

Fleming, L. & Waguespack, D. (2007). Brokerage, boundary spanning, and leadership in open innovation communities, *Organization Science* **18**(2): 165–180.

Füller, J. (2006). Why consumers engage in virtual new product developments initiated by producers, *Advances in Consumer Research* **33**(1): 639–646.

Fogel, K., Barkhau, M., Menge, S. & Pittinger, R. (2010). *Produktion von Open-Source-Software: Wie man ein erfolgreiches freies Software-Projekt führt*, Published under Creative Commons License, URL, http://producingoss.com/de/index.html.

Fosfuri, A. (2006). The licensing dilemma: Understanding the determinants of the rate of technology licensing, *Strategic Management Journal* **27**: 1141–1158.

Fosfuri, A., Giarratana, M. & Luzzi, A. (2008). The penguin has entered the building: The commercialization of open source software products, *Organization Science* **19**(2): 292–305.

Foss, K. & Foss, N. (2005). Resources and transaction costs: How property rights economics furthers the resource-based view, *Strategic Management Journal* **26**(2): 541–553.

Foss, N. (1996). Knowledge-based approaches to the theory of the firm: Some critical comments, *Organization Science* **7**(5): 470–476.

Foss, N., Laursen, K. & Pedersen, T. (2011). Linking customer interaction and innovation: The mediating role of new organizational practices, *Organization Science* **22**(4): 980–999.

Franck, E. & Jungwirth, C. (2003a). Die Governance von Open Source Projekten, *Zeitschfift für Betriebswirtschaft (ZfB)* **73**: 1–21.

Franck, E. & Jungwirth, C. (2003b). Reconciling investors and donators – The governance structure of open source, *Journal of Management and Governance* **7**(4): 401–421.

Frank, U. (2006). Towards a pluralistic conception of research methods in information systems, *ICB Research Report No. 7, University of Duisburg-Essen* .

Frank, U., Schauer, C. & Wigand, R. (2008). Different paths of development of two information systems communities: A comparative study based on peer interviews, *Communications of the AIS* **22**(1): 391–412.

Franke, N. & Shah, S. (2003). How communities support innovative activities: An exploration of assistance and sharing among end-users, *Research Policy* **32**: 157–178.

Freedman, J. & Fraser, S. (1966). Complience without pressure: The foot-in-the-door technique, *Journal of Personality and Social Psychology* **4**(2): 195–202.

Freeman, L. (1979). Centrality in social networks, *Social Networks* **1**: 585–592.

Frese, E. (2000). *Grundlagen der Organisation: Konzept - Prinzipien - Strukturen*, 7th edn, Gabler, Wiesbaden, Germany.

Friedman, R. & Podolny, J. (1992). Differentiation of boundary spanning roles: Labor negotiations and implications for role conflict, *Administrative Science Quarterly* **37**: 28–47.

Galbraith, J. (1973). *Designing Complex Organizations*, Addison-Wesley, Reading, MA.

Gallaugher, J. & Wang, Y.-M. (2002). Understanding network effects in software markets: Evidence from web server pricing, *MIS Quarterly* **26**(4): 303–327.

Gambardella, A. & Hall, B. (2006). Proprietary vs public domain licensing of software and research products, *Research Policy* **35**: 875–892.

Gambardella, A. & McGahan, A. (2010). Business model innovation: General purpose technologies and their implications for industry structure, *Long Range Planning* **43**(2/3): 262–271.

Gamma, E., Helm, R., Johnson, R. & Vlissides, J. (1995). *Design Patterns: Elements of Reusable Object-Oriented Software*, Addison-Wesley, Boston, MA.

Garringa, H., Von Krogh, G. & Spaeth, S. (2011). Joining forces: A framework of multi-partner alliances for public good innovations, *Paper, presented at 11th European Academy of Management Conference (EURAM)*, Tallinn, Estonia.

Garzarelli, G. & Fontanella, R. (2011). Open source software production, spontaneous input, and organizational learning, *American Journal of Economics and Sociology* **70**(4): 928–950.

Gassmann, O. & Enkel, E. (2004). Towards a theory of open innovation: Three core process archetypes, *Proceedings of the R&D Management Conference (RADMA)*, Lisbon, Portugal.

Gassmann, O., Kausch, C. & Enkel, E. (2010). Negative side effects of customer integration, *International Journal of Technology Management* **50**(1): 43–62.

Gassmann, O., Sandmeier, P. & Wecht, C. (2006). Extreme customer integration in the front-end: Learning from a new software paradigm, *International Journal of Technology Management* **33**(1): 46–66.

Gavetti, G. & Rivkin, J. (2007). On the origin of strategy: Action and cognition over time, *Organization Science* **18**(3): 420–439.

Ghanpanchi, A., Aurum, A. & Low, G. (2011). A taxonomy for measuring the success of open source projects, *First Monday* **16**(8).

Ghosh, R. (2005). Understanding free software developers: Findings from the FLOSS study, *in* J. Feller, B. Fitzgerald, S. Hissam & K. Lakhani (eds), *Perspectives on Free and Open Source Software*, MIT Press, Cambridge, Mass, pp. 23–45.

Ghosh, R. & David, P. (2003). The nature and composition of the linux kernel developer community: A dynamic analysis, URL, http://dxm.org/papers/licks1/licksresults.pdf, last access: 8/30/2011.

Giddens, A. (1979). *Central Problems in Social Theory: Action, Structure and Contradiction in Social Analysis*, University of California Press, Berkeley / Los Angeles, CA.

Giddens, A. (1984). *The Constitution of Society: Outline of the Theory of Structuration*, University of California Press, Berkeley / Los Angeles, CA.

Giuri, P., Ploner, M., Rullani, F. & Torrisi, S. (2010). Skills, division of labor and performance in collective inventions: Evidence from open source software, *International Journal of Industrial Organization* **28**(1): 54–68.

Giuri, P., Rullani, F. & Torrisi, S. (2008). Explaining leadership in open source software projects, *Information Economics and Policy* **20**: 305–315.

Goldman, R. & Gabriel, R. (2005). *Innovation Happens Elsewhere: Open Source as Business Strategy*, Morgan Kaufmann, Amsterdam, NL.

Gong, Y., Huang, J.-C. & Farh, J.-L. (2009). Employee learning orientation, transformational leadership, and employee creativity: The mediating role of employee creative self-efficacy, *Academy of Management Journal* **54**(4): 765–778.

Grand, S., von Krogh, G., Leonard, D. & Swap, W. (2004). Resource allocation beyond firm boundaries: A multi-level model for open source innovation, *Long Range Planning* **37**: 591–610.

Granovetter, M. (1973). The strength of weak ties, *American Journal of Sociology* **78**(5): 1360–1380.

Granovetter, M. (1983). The strength of weak ties: A network theory revisited, *Sociological Theory* **1**: 201–233.

Grant, R. (1996a). Prospering in dynamically-competitive environments: Organizational capability as knowledge integration, *Organization Science* **7**(4): 375–387.

Grant, R. (1996b). Toward a knowledge-based theory of the firm, *Strategic Management Journal* **17**: 109–122.

Green, J. & Stokey, N. (1983). A comparison of tournaments and contracts, *Journal of Political Economy* **91**(3): 349–364.

Greene, W. (2000). *Econometric Analysis*, 4th edn, Prentice Hall, Upper Saddly River, NJ.

Grewal, R., Lilien, G. & Mallapragada, G. (2006). Location, location, location: How network embeddedness affects project success in open source systems, *Management Science* **52**(7): 1043–1056.

Gruber, M. & Henkel, J. (2006). New ventures based on open innovation: An emprical analysis of start-up firms in embedded Linux, *International Journal of Technology Management* **33**: 356–372.

Gutenberg, E. (1983). *Grundlagen der Betriebswirtschaftslehre - Band 1: Die Produktion*, Springer, Berlin, Germany.

Hagedoorn, J., Cloodt, D. & Van Kranenburg, H. (2005). Intellectual property rights and the governance of international R&D partnerships, *Journal of International Business Studies* **36**(2): 175–186.

Hahn, J., Moon, J. & Zhang, C. (2008). Emergence of new project teams from open source software developer networks: Impact of prior collaboration ties, *Information Systems Research* **19**(3): 369–391.

Hair, J., Anderson, R., Tatham, R. & Black, W. (1998). *Multivariate Data Analysis*, 5th edn, Prentice Hall, Englewood Cliffs, NJ.

Hampe, J. F. (2010). From mobile service research to market success: Some considerations on bridging gaps between academia and industry, Inaugural speech for KPN-founded Cor Wit endowed chair, Delft University of Technology, May 27th, 2010, Delft, The Netherlands.

Hansen, M. (1999). The search-transfer problem: The role of weak ties in sharing knowledge across organization subunits, *Administrative Science Quarterly* **44**: 82–111.

Hansen, M. (2002). Knowledge networks: Explaining effective knowledge sharing in multiunit companies, *Organization Studies* **13**: 222–248.

Hansen, M., Mors, M. & Løvås, B. (2005). Knowledge sharing in organizations: Multiple networks, multiple phases, *Academy of Management Journal* **48**(5): 776–793.

Harhoff, D., Henkel, J. & Von Hippel, E. (2003). Profiting from voluntary information spillovers: How users benefit by freely revealing their innovations, *Research Policy* **32**: 1753–1769.

Harhoff, D. & Mayrhofer, P. (2010). Managing user communities and hybrid innovation processes: Concepts and design implications, *Organizational Dynamics* **39**(2): 137–144.

Harrison, D. & Waluszewski, A. (2008). The development of a user-network as a way to re-launch an unwanted product, *Research Policy* **37**: 115–130.

Hars, A. & Ou, S. (2002). Working for free? Motivations for participation in open source projects, *International Journal of Electronic Commerce* **6**(3): 25–39.

Hauschildt, J. & Chakrabarti, A. (1989). The division of labour in innovation management, *R&D Management* **19**(2): 161–171.

Hayes, R., Wheelwright, S. & Clark, K. (1988). *Dynamic Manufacturing: Creating the Learning Organization*, Free Press, New York, NY.

He, Z.-L. & Wong, P.-K. (2004). Exploration vs. exploitation: An empirical test of the ambidexterity hypothesis, *Organization Science* **15**(4): 481–494.

Hecker, F. (1999). Setting up shop: The business of open-source software, *IEEE Software* **16**(1): 45–51.

Heckman, R., Crowston, K., Li, Q., Allen, E., Eseryel, U. Y., Howison, J. & Wei, K. (2006). Emergent decision-making practices in technology-supported self-organizing distributed teams, *Proceedings of the International Conference on Information Systems (ICIS 2006)*, Milwaukee, WI.

Hellmann, T. (2006). IPOs, acquisitions, and the use of convertible securities in venture capital, *Journal of Financial Economics* **81**: 649–679.

Hellmann, T. & Puri, M. (2002). Venture capital and the professionalization of start-up firms: Empirical evidence, *Journal of Finance* **57**(1): 169–198.

Helpman, E. (1993). Innovation, imitation, and intellectual property rights, *Econometrica* **61**(6): 1247–1280.

Hemphill, T. (2006). A taxonomy of closed and open source software industry business models, *International Journal of Innovation & Technology Management* **3**(1): 61–82.

Henderson, R. & Clark, K. (1990). Architectural innovation: The reconfiguration of existing product technologies and the failure of established firms, *Administrative Science Quarterly* **44**(1): 9–30.

Henkel, J. (2006). Selective revealing in open innovation processes: The case of embedded Linux, *Research Policy* **35**(7): 953–969.

Henkel, J. (2007). *Offene Innovationsprozesse: Die kommerzielle Entwicklung von Open Source Software*, Deutscher Universitäts-Verlag, Wiesbaden, Germany.

Henkel, J. (2009). Champions of revealing: The role of open source developers in commercial firms, *Industrial and Corporate Change* **18**(3): 435–471.

Henry, J. (1999). John Locke, property rights, and economic theory, *Journal of Economic Issues* **33**(3): 609–624.

Hertel, G. (2007). Motivating job design as a factor in open source governance, *Journal of Management and Governance* **11**: 129–137.

Hertel, G., Niedner, S. & Herrmann, S. (2003). Motivation of software developers in open source projects: An internet-based survey of contributors to the Linux kernel, *Research Policy* **32**: 1159–1177.

Hevner, A., March, S., Park, J. & Ram, S. (2004). Design science in information systems research, *MIS Quarterly* **28**(1): 75–105.

Holmström, B. (1979). Moral hazard and observability, *The Bell Journal of Economics* **10**(1): 74–91.

Homans, G. (1958). Social behavior as exchange, *American Journal of Sociology* **63**: 597–606.

Hosmer, D. & Lemeshow, S. (2000). *Applied Logistic Regression*, John Wiley & Sons, New York, NY.

Hu, D. & Zhao, J. (2009). Discovering determinants of project participation in an open source social network, *Proceedings of the International Conference on Information Systems (ICIS 2009)*, Phoenix, AZ.

Huang, F. & Rice, J. (2009). The role of absorptive capacity in faciliating "open innovation" outcomes: A study of Australian SMEs in the manufacturing sector, *International Journal of Innovation Management* **13**(2): 201–220.

Hunt, S. (1983). *Marketing Theory – The Philosophy of Marketing Science*, Irwin, Homewood, IL.

Hutter, K., Hautz, J., Füller, J., Mueller, J. & Matzler, K. (2011). Communition: The tension between competition and collaboration in community-based design contests, *Creativity and Innovation Management* **20**(1): 3–21.

Iannacci, F. (2005). Coordination processes in open source software development: The Linux case study, *Emergence: Complexity and Organization* **7**(2): 20–30.

Iannacci, F. & Mitleton-Kelly, E. (2005). Beyond markets and firms: The emergence of open source networks, *First Monday* **10**(5).

Iansiti, M. & Richards, G. (2006). The business of free software: Enterprise incentives, investment and motivation in the open source community, *Harvard Business School Working Paper Series, No. 07-028*
.

Ibarra, H. (1993). Network, centrality, power, and innovation involvement: Determinants of technical and administrative roles, *Academy of Management Journal* **36**: 471–501.

Ibarra, H. (2004). *Working Identity: Unconventional Strategies for Reinventing your Career*, Harvard Business School Press, Cambridge, MA.

Ives, B. & Olson, M. (1984). User involvement and MIS success: A review of research, *Management Science* **30**(5): 586–603.

Jaffe, A. (1986). Technological opportunity and spillovers of R&D: Evidence from firms' patents, profits and market value, *American Economic Review* **76**: 984–1001.

Jaworski, B. (1988). Toward a theory of marketing control: Environmental context, control types, and consequences, *Journal of Marketing* **52**(3): 23–39.

Jensen, M., Johnson, B., Lorenz, E. & Lundvall, B. (2007). Forms of knowledge and modes of innovation, *Research Policy* **36**: 680–693.

Jensen, M. & Meckling, W. (1976). Theory of the firm. managerial behavior, agency costs, and ownership structure, *Journal of Financial Economics* **3**(4): 305–360.

Jones, C., Hesterly, W. & Borgatti, S. (1997). A general theory of network governance: Exchange conditions and social mechanisms, *Academy of Management Review* **22**(4): 911–945.

Jones, G., Lanctot, A. & Teegen, H. (2001). Determinants and performance impacts of external technology acquisition, *Journal of Business Venturing* **16**(3): 255–283.

Jullien, N. & Zimmermann, J. (2009). Firms' contribution to open-source software and the dominant user's skill, *European Management Review* **6**(2): 130–139.

Katz, M. & Shapiro, C. (1986). Technology adoption in the presence of network externalities, *Journal of Political Economy* **94**(4): 822–841.

Katz, R. & Allen, T. (1982). Investigating the not-invented-here (NIH) syndrome: A look at performance, tenure and communication patterns of 50 R&D project groups, *R&D Management* **12**: 7–19.

Keßler, S. & Alpar, P. (2009). Customization of open source software in companies, *Proceedings of the 5th International Conference on Open Source Systems (OSS)*, Springer, Skovde, Sweden, pp. 129–142.

Kellogg, K., Orlikowski, W. J. & Yates, J. (2006). Life in the trading zone: Structuring coordination across boundaries in postbureaucratic organizations, *Organization Science* **17**: 22–44.

Kennedy, D. (2004). What lawyers need to know about the open source licenses, *Journal of Internet Law* **7**: 3–10.

Keupp, M. & Gassmann, O. (2009). Determinants and archetypes of users of open innovation, *R&D Management* **39**(4): 331–341.

Khong, D. (2006). The historical law and economics of the first copyright act author, *Erasmus Law and Economics Review* **2**(1): 35–69.

Kieser, A. & Walgenbach, P. (2010). *Organisation*, 6th edn, Schäffer-Poeschel, Stuttgart, Germany.

Kijkuit, B. & Van den Ende, J. (2007). The organizational life of an idea: Integrating social network, creativity and decision making, *Journal of Management Studies* **44**: 863–882.

Killduff, M. & Tsai, W. (2003). *Social Networks and Organizations*, Sage Publications, London, U.K. / Thousand Oaks, CA.

Kim, A. J. (2000). *Community Building on the Web: Secret Strategies for Successful Online Communities*, Addison Wesley, London, U.K.

Kirsch, L. (1997). Portfolios of control modes and is project management, *Information Systems Research* **8**(3): 215–238.

Kirsch, L., Ko, D.-G. & Haney, M. (2010). Investing the antecedents of team-based clan control: Adding social capital as a predictor, *Organization Science* **21**(2): 469–489.

Kleinaltenkamp, M. (2002). Customer integration im electronic business, *in* R. Weiber (ed.), *Handbuch Electronic Business*, 2nd edn, Gabler, Wiesbaden, Germany, pp. 443–468.

Kleinberg, J., Suri, S., Tardos, E. & Wexler, T. (2008). Strategic network formation with structural holes, *Proceedings of the 9th ACM Conference on Electronic Commerce*, Chicago, IL.

Knox, H., Savage, M. & Harvey, P. (2006). Social networks and the study of relations: Networks as method, metaphor and form, *Economy and Society* **35**(1): 113–140.

Koch, S. & Schneider, G. (2002). Effort, cooperation and coordination in an open source software project: Gnome, *Information Systems Journal* **12**: 27–42.

Koglin, O. & Metzger, A. (2004). Urheber- und Lizenzrecht im Bereich von Open-Source-Software, *in* R. Gehring & B. Lutterbeck (eds), *Open-Source-Jahrbuch 2004*, Lehmanns Media, Berlin, Germany.

Kogut, B. & Metiu, A. (2001). Open source software development and distributed innovation, *Oxford Review of Economic Policy* **17**: 248–264.

Kogut, B. & Zander, U. (1992). Knowledge of the firm, combinative capabilities, and the replication of technology, *Organization Science* **3**(3): 383–397.

Kohli, R. & Kettinger, W. (2004). Informating the clan: Controlling physicians' costs and outcomes, *MIS Quarterly* **28**(3): 363–394.

Koski, H. (2005). OSS production and licensing strategies of software firms, *Review of Economic Research on Copyright Issues* **2**(2): 111–125.

Kotlarsky, J. & Oshri, I. (2005). Social ties, knowledge sharing and successful collaboration in globally distributed system development projects, *European Journal of Information Systems* **14**(1): 37–48.

Krackhart, D. (1993). Graph theoretical dimensions of informal organizations, *in* K. Carley & M. Prietula (eds), *Computational Organization Theory*, Lawrence Erlbaum, Hillsdale, NJ, pp. 89–111.

Krafft, M. (1999). An empirical investigation of the antecedents of sales force control systems, *Journal of Marketing* **63**: 120–134.

Krishnamurthy, S. (2003). A managerial overview of open source software, *Business Horizons* **46**(5): 47–56.

Krishnamurthy, S. (2005). An Analysis of Open Source Business Models, *in* J. Feller, B. Fitzgerald, S. Hissam & K. Lakhani (eds), *Perspectives on Free and Open Source Software*, MIT Press, Cambridge, Mass, pp. 279–296.

Krishnamurthy, S. (2006). On the intrinsic and extrinsic motivation of free/libre/open source (FLOSS) developers, *Knowledge, Technology and Policy* **18**(4): 17–39.

Krishnamurthy, S. & Tripathi, A. (2009). Monetary donations to an open source software platform, *Research Policy* **38**: 404–414.

Kroah-Hartmann, G., Corbet, J. & McPherson, A. (2008). Linux kernel development (April 2008), *Technical report*, Linux Foundation, San Francisco, CA.

Krumke, S. & Noltemeier, H. (2005). *Graphentheoretische Konzepte und Algorithmen*, Teubner Verlag, Wiesbaden, Germany.

Kuhn, T. (1962). *The Structure of Scientific Revolutions*, Chicago University Press, Chicago, IL.

Kuk, G. (2006). Strategic interaction and knowledge sharing in the KDE developer mailing list, *Management Science* **52**(7): 1031–1042.

Lakhani, K. (2006). *The Core and the Peripherie in Distributed and Self-Organizing Systems*, PhD thesis, MIT Sloan School of Management, Cambridge, MA.

Lakhani, K. & Von Hippel, E. (2003). How open source software works: Free user-to-user assistance, *Research Policy* **32**: 923–943.

Lakhani, K. & Wolf, R. (2005). Why hackers do what they do: Understanding motivation and effort in free/open source software projects, *in* J. Feller, B. Fitzgerald, S. Hissam & K. Lakhani (eds), *Perspectives on Free and Open Source Software*, MIT Press, Cambridge, Mass, pp. 3–22.

Lakka, S., Stamati, T., Michalakelis, C. & Martakos, D. (2011). The ontology of the OSS business model: An exploratory study, *International Journal of Open Source Software & Processes* **3**(1): 39–59.

Lambe, C. & Spekman, R. (1997). Alliances, external technology acquisition, and discontinuous technological change, *Journal of Product Innovation Management* **14**(2): 102–116.

Lane, P. J., Koka, B. & Pathak, S. (2006). The reification of absorptive capacity: A critical review and rejuvenation of the construct, *Academy of Management Review* **31**: 833–863.

Langlois, R. (2002). Modularity in technology and organization, *Journal of Economic Behavior and Organization* **49**: 19–37.

Langlois, R. & Garzarelli, G. (2008). Of hackers and hairdressers: Modularity and the organizational economics of open-source collaboration, *Industry and Innovation* **15**: 125–143.

Lattemann, C. & Stieglitz, S. (2005). Framework for governance in open source communities, *Proceedings of the 38th Hawaii International Conference on System Sciences (HICSS-38)*, Manoa, Hawaii.

Laursen, K. & Salter, A. (2006). Open for innovation: The role of openness in explaining innovation performance among U.K. manufacturing firms, *Strategic Management Journal* **27**: 131–150.

Lazer, D. & Friedman, A. (2007). The network structure of exploration and exploitation, *Administrative Science Quarterly* **52**(4): 667–694.

Leavitt, H. (1951). Some effects of certain communication patterns on group performance, *Journal of Abnormal and Social Psychology* **46**: 38–50.

Lee, G. & Cole, R. (2003). From a firm-based to a community-based model of knowledge creation: The case of the Linux kernel development, *Organization Science* **14**(6): 633–649.

Lee, S., Kim, H.-W. & Gupta, S. (2009). Measuring open source software success, *Omega* **37**(2): 426–438.

Leijonhufvud, A. (1986). Capitalism and the factory system, *in* R. N. Langlois (ed.), *Economics as a Process: Essays in the New Institutional Economics*, Cambridge University Press, New York, NY, pp. 203–223.

Lepak, D., Smith, K. & Taylor, S. (2007). Value creation and value capture: A multilevel perspective, *Academy of Management Review* **32**(1): 180–194.

Lerner, J. & Schankerman, M. (2010). *The Comingled Code. Open Source and Economic Development*, MIT Press, Cambridge, MA.

Lerner, J. & Tirole, J. (2002). Some simple economics of open source, *Journal of Industrial Economics* **50**(2): 197–234.

Levinthal, D. & Wu, B. (2010). Opportunity costs and non-scale free capabilities: Profit maximization, corporate scope, and profit margins, *Strategic Management Journal* **31**: 780–801.

Lewin, P. (2007). Creativity or coercion: Alternative perspectives on rights to intellectual property, *Journal of Business Ethics* **71**(4): 441–455.

Li, M. & Gao, F. (2003). Why Nonaka highlights tacit knowledge: A critical review, *Journal of Knowledge Management* **7**(4): 6–14.

Lichtenthaler, U. (2005). External commercialization of knowledge: Review and research agenda, *International Journal of Management Reviews* **7**: 231–255.

Lichtenthaler, U. (2007). Understanding the determinants of external technology commercialization, *ZfB - Zeitschrift für Betriebswirtschaft* **4**(Special Issue): 21–45.

Lichtenthaler, U. (2009a). Absorptive capacity, environmental turbulence, and the complementarity of organizational learning processes, *Academy of Management Journal* **52**(4): 822–846.

Lichtenthaler, U. (2009b). Outbound open innovation and its effect on firm performance: Examining environmental influences, *R&D Management* **39**(4): 317–330.

Lichtenthaler, U. (2010a). Intellectual property and open innovation: An empirical analysis, *International Journal of Technology Management* **52**(3/4): 372–391.

Lichtenthaler, U. (2010b). Outward knowledge transfer: The impact of project-based organization on performance, *Industrial and Corporate Change* **19**(6): 1705–1739.

Lichtenthaler, U. (2010c). Technology exploitation in the context of open innovation: Finding the right "job" for your technology, *Technovation* **30**(7/8): 429–435.

Lichtenthaler, U. (2011a). The evolution of technology licensing management: Identifying five strategic approaches, *R&D Management* **41**(2): 173–189.

Lichtenthaler, U. (2011b). Is open innovation a field of study or a communication barrier to theory development? A contribution to the current debate, *Technovation* **31**(2-3): 138–139.

Lichtenthaler, U. (2011c). Open innovation: Past research, current debates, and future directions, *Academy of Management Perspectives* **25**(1): 75–93.

Lichtenthaler, U. & Ernst, H. (2006). Attitudes to externally organising management tasks: A review, reconsideration and extension of the NIH syndrome, *R&D Management* **36**(4): 367–386.

Lichtenthaler, U. & Ernst, H. (2007). External technology commercialization in large firms: Results of a quantitative benchmark study, *R&D Management* **37**(5): 383–397.

Lichtenthaler, U. & Ernst, H. (2009a). Opening up the innovation process: The role of technology aggressiveness, *R&D Management* **39**(1): 38–54.

Lichtenthaler, U. & Ernst, H. (2009b). The role of champions in the external commercialization of knowledge, *Journal of Product Innovation Management* **26**: 371–387.

Lichtenthaler, U., Ernst, H. & Conley, J. (2011). How to develop a successful technology licensing program, *MIT Sloan Management Review* **52**(2): 17–19.

Lichtenthaler, U., Ernst, H. & Hoegl, M. (2010). Not-sold-here: How attitudes influence external knowledge exploitation, *Organization Science* **21**(5): 1054–1071.

Lichtenthaler, U. & Lichtenthaler, E. (2004). Organisation of international external technology acquisition projects, *International Journal of Technology Transfer and Commercialisation* **3**(3): 291–307.

Lichtenthaler, U. & Lichtenthaler, E. (2009). A capability-based framework for open innovation: Complementing absorptive capacity, *Journal of Management Studies* **46**(8): 1315–1338.

Lichtenthaler, U. & Lichtenthaler, E. (2010). Technology transfer across organizational boundaries: Absorptive capacity and desorptive capacity, *California Management Review* **53**(1): 154–170.

Liebowitz, S. & Margolis, S. (1994). Network externality: An uncommon tragedy, *Journal of Economic Perspectives* **8**(2): 133–150.

Lindgren, M. & Packendorff, J. (2011). Issues, responsibilities and identities: A distributed leadership perspective on biotechnology R&D management, *Creativity and Innovation Management* **20**(3): 157–170.

Lindman, J., Rossi, M. & Puustell, A. (2011). Matching open source software licenses with corresponding business models, *IEEE Software* **28**(4): 31–35.

Long Lingo, E. & O'Mahony, S. (2010). Nexus work: Brokerage on creative projects, *Administrative Science Quarterly* **55**: 47–81.

Long, Y. & Siau, K. (2008). Impacts of social network structure on knowledge sharing in open source development teams, *Proceedings of the 14th Americas Conference of Information Systems (AMCIS)*, Toronto, Canada.

López, L. & Roberts, E. (2002). First-mover advantages in regimes of weak appropriability: The case of financial services innovations, *Journal of Business Research* **55**: 997–1005.

Lüthje, C. & Herstatt, C. (2004). The lead user method: An outline of empirical findings and issues for future research, *R&D Management* **34**(5): 553–568.

Lukach, R., Kort, P. & Plasmans, J. (2007). Optimal R&D investment strategies under the threat of new technology entry, *International Journal of Industrial Organization* **25**: 103–119.

MacCormack, A., Baldwin, C. & Rusnak, J. (2010). The architecture of complex design: Do core-periphery structures dominate?, *Harvard Business School Working Paper Series, No. 10-059* .

MacCormack, A. & Iansiti, M. (2009). Intellectual property, architecture and the management of technological transitions: Evidence from microsoft corporation, *Journal of Product Innovation Management* **26**: 248–263.

MacCormack, A., Rusnak, J. & Baldwin, C. (2006). Exploring the structure of complex software designs: An empirical study of open source and proprietary code, *Management Science* **52**(7): 1015–1030.

MacCormack, A., Rusnak, J. & Baldwin, C. (2008). Exploring the duality between product and organizational architectures: A test of the mirroring hypothesis, *Harvard Business School Working Paper Series, No. 08-039* .

Magretta, J. (2002). Why business models matter, *Harvard Business Review* **80**(5).

Mahadevan, B. (2000). Business models for internet-based e-commerce: An anatomy, *California Management Review* **42**(4): 55–69.

Makadok, R. (2001). Toward a synthesis of the resource-based and dynamic capabilities view of rent creation, *Strategic Management Journal* **22**: 387–401.

Mann, R. (2006). Commercializing open source software: Do property rights still matter?, *Harvard J. Law Technology* **20**: 1–46.

Mansfield, E. (1986). Patents and innovation: An empirical study, *Management Science* **32**(2): 173–181.

March, J. (1991). Exploration and exploitation in organizational learning, *Organization Science* **2**: 71–87.

March, S. & Smith, G. (1995). Design and natural science research on information technology, *Decision Support Systems* **15**: 251–266.

Markus, M. (2007). The governance of free/open source software projects: monlithic, multidimensional, or configurational?, *Journal of Management and Governance* **11**(2): 151–163.

Marple, D. (1961). The decisions of engineering design, *IEEE Transactions of Engineering Management* **2**: 55–71.

Marsden, P. (1990). Network data and measurement, *Annual Review of Sociology* **16**: 435–463.

Martins, E. & Terblanche, F. (2003). Building organizational culture that stimulates creativity and innovation, *European Journal of Innovation Management* **6**(1): 64–74.

Masmoudi, H., den Besten, M., de Loupy, C. & Dalle, J.-M. (2009). Peeling the onion: The words and actions that distinguish core from periphery in bug reports and how core and periphery act together, *Proceedings of the 5th International Conference on Open Source Systems (OSS)*, Springer, Skovde, Sweden, pp. 284–297.

Mata, F., Fuerst, W. & Barney, J. (1995). Information technology and sustained competitive advantage: A resource based analysis, *MIS Quarterly* **19**(4): 487–505.

Mathews, J. (2003). Competitive dynamics and economic learning: An extended resource based view, *Industrial and Corporate Change* **12**(1): 115–145.

McCloskey, D. (1985a). The loss function has been mislaid: The rhetoric of significant tests, *The American Economic Review* **75**(2): 201–205.

McCloskey, D. (1985b). *The Rhetoric of Economics*, University of Wisconsin Press, Madison, WS.

McCloskey, D. & Ziliak, S. (1996). The standard error of regression, *Journal of Economic Literature* **34**(1): 97–114.

McGowan, D. (2005). Legal aspects of free and open source software, *in* J. Feller, B. Fitzgerald, S. Hissam & K. Lakhani (eds), *Perspectives on Free and Open Source Software*, MIT Press, Cambridge, Mass, pp. 361–391.

McMahon, J. & Ivancevich, J. (1976). A study of control in manufacturing organization: Managers and nonmanagers, *Administrative Science Quarterly* **21**(1): 66–83.

Mehra, A., Dewan, R. & Freimer, M. (2011). Firms as incubators of open source software, *Information Systems Research* **22**: 22–38.

Merges, R. (2004). A new dynamism in the public domain, *University of Chicago Law Review* **71**(1): 183–203.

Milev, R., Muegge, S. & Weiss, M. (2009). Design evolution of an open source project using an improved modularity metric, *Open Source Ecosystems: Diverse Communities Interacting, Proceedings of the 5th OSS Conference, Skovde, Sweden*, Springer, pp. 20–33.

Milgrom, P. & Roberts, J. (1992). *Economics, Organization, and Management*, Prentice-Hall, Englewood Cliffs, NJ.

Miller, D., Fern, M. & Cardinal, L. (2007). The use of knowledge for technological innovation within diversified firms, *Academy of Management Journal* **50**(2): 308–326.

Mir, R. & Watson, A. (2000). Strategic management and the philosophy of science: The case for a constructivist methodology, *Strategic Management Journal* **21**(9): 941–953.

Méndez-Durón, R. & Garcia, C. (2009). Returns from social capital in open source software networks, *Journal of Evolutionary Economics* **19**: 277–295.

Mockus, A., Fielding, R. & Herbsleb, J. (2002). Two case studies of open source software development: Apache and Mozilla, *ACM Transactions on Software Engineering and Methodology* **11**(3): 309–346.

Mol, J. & Wijnberg, N. (2011). From resources to value and back: Competition between and within organizations, *British Journal of Management* **22**(1): 77–95.

Moon, J. & Sproull, L. (2002). Essence of distributed work: The case of the Linux kernel, *Distributed Work*, MIT Press, Cambridge, Mass., pp. 381–404.

Morgan, L. & Finnegan, P. (2008). Deciding on open innovation: An exploration of how firms create and capture value with open source software, *Open IT-Based Innovation: Moving towards coopertaive IT transfer and knowledge diffusion*, Springer, Boston, Mass., pp. 229–246.

Mosakowski, E. (1998). Entrepreneurial resources, organizational choices, and competitive outcomes, *Organization Science* **9**(6): 625–643.

Mowery, D., Oxley, J. & Silverman, B. (1996). Strategic alliances and interfirm knowledge transfer, *Strategic Management Journal* **17**: 77–91.

Muller, P. (2006). Reputation, trust and the dynamics of leadership in communities of practice, *Journal of Governance and Management* **10**(4): 381–400.

Nahapiet, J. & Ghoshal, S. (1998). Social capital, intellectual capital, and the organizational advantage, *Academy of Management Review* **23**: 242–266.

Nambisan, P. & Nambisan, S. (2009). Models of consumer value cocreation in health care, *Health Care Management Review* **34**(4): 344–354.

Nambisan, S. (2002). Designing virtual customer environments for new product development: Toward a theory, *Academy of Management Review* **27**(3): 392–413.

Nambisan, S. & Baron, R. (2009). Virtual customer environments: Testing a model of voluntary participation in value co-creation activities, *Journal of Product Innovation Management* **26**(4): 388–406.

Narayanan, V., Yang, Y. & Zahra, S. (2009). Corporate venturing and value creation: A review and proposed framework, *Research Policy* **38**: 58–76.

Newcomer, E. & Lomow, G. (2005). *Understanding SOA with Web Services*, Addison Wesley, Upper Saddle River, NJ.

Neyer, A.-K., Bullinger, A. & Moeslein, K. (2009). Integrating inside and outside innovators: A sociotechnical systems perspective, *R&D Management* **39**: 410–419.

Niehaves, B. (2005). Epistemological perspectives on multi-method information systems research, *Proceedings of the 13th European Conference on Information Systems (ECIS)*, Regensburg, Germany.

Nonaka, I. & Takeuchi, H. (1995). *The Knowledge Creating Company*, Oxford University Press, New York, NY.

Nunamaker, J., Chen, M. & Purdin, T. (2004). Systems development in informations systems research, *Journal of Management Information Systems* **7**(3): 89–106.

Oesterle, H., Becker, J., Frank, U., Hess, T., Karagiannis, D., Krcmar, H., Loos, P., Mertens, P., Oberweis, P. & Sinz, E. (2010). Memorandum on design-oriented information systems research, *European Journal of Information Systems* **20**: 7–20.

O'Hern, M. & Rindfleisch, A. (2010). Customer co-creation: A typology and research agenda, *Review of Marketing Research* **6**: 84–106.

Okhuysen, G. & Bechky, B. (2009). Coordination in organizations: An integrative perspective, *Annals of Academy of Management* **3**(1): 463–502.

Oliveira, P. & Von Hippel, E. (2011). Users as service innovators: The case of banking services, *Research Policy* **40**(6): 806–818.

Olson, E. & Bakke, G. (2001). Implementing the lead user method in a high technology firm: A longitudinal study of intentions versus actions, *Journal of Product Innovation Management* **18**(6): 388–395.

Olson, M. (2005). Dual licensing, *Open Sources 2.0: The Continuing Evolution*, O'Reilly, Sebastopol, CA, pp. 93–107.

O'Mahony, M. & Vecchi, M. (2009). R&D, knowledge spillovers and company productivity performance, *Research Policy* **38**: 35–44.

O'Mahony, S. (2003). Guarding the commons: How community managed software projects protect their work, *Research Policy* **32**(7): 1179–1198.

O'Mahony, S. (2005). Nonprofit foundations and their role in community-firm software collaboration, *in* J. Feller, B. Fitzgerald, S. Hissam & K. Lakhani (eds), *Perspectives on Free and Open Source Software*, MIT Press, Cambridge, Mass, pp. 393–414.

O'Mahony, S. (2007). The governance of open source initiatives: What does it mean to be community managed?, *Journal of Management and Governance* **11**(2): 139–150.

O'Mahony, S. & Bechky, B. (2006). Stretchwork: Managing the career progression paradox in external labor markets, *Academy of Management Journal* **49**(5): 917–941.

O'Mahony, S. & Bechky, B. (2008). Boundary organizations: Enabling collaboration between unexpected allies, *Administrative Science Quarterly* **53**: 422–459.

O'Mahony, S., Cela Diaz, F. & Mamas, E. (2005). IBM and Eclipse, *Harvard Business School Teaching Case, No. 9-906-007*.

O'Mahony, S. & Ferraro, F. (2007). The emergence of governance in an open source community, *Academy of Management Journal* **50**(5): 1079–1106.

O'Reilly III., C. A. & Tushman, M. L. (2011). Organizational ambidexterity in action: How managers explore and exploit, *California Management Review* **53**(4): 5–22.

Orlikowski, W. & Baroudi, J. (1991). Studying information technology in organizations: Research approaches and assumptions, *Information Systems Research* **2**(1): 1–28.

Osterloh, M. & Rota, S. (2007). Open source development – Just another case of collective invention?, *Research Policy* **36**: 157–171.

Osterwalder, A. & Pigneur, Y. (2010). *Business Model Generation: A Handbook for Visionaries, Game Changers, and Challengers*, John Wiley & Sons, New York, NY.

Osterwalder, A., Pigneur, Y. & Tucci, C. (2005). Clarifying business models: Origins, present, and future of the concept, *Communications of the AIS* **16**: 1–38.

Ouchi, W. (1977). The relationship between organizational structure and organizational control, *Administrative Science Quarterly* **22**(1): 95–113.

Ouchi, W. (1979). A conceptual framework for the design of organizational control mechanisms, *Management Science* **25**: 838–848.

Ouchi, W. (1980). Markets, bureaucracies, and clans, *Administrative Science Quarterly* **25**: 129–141.

Ouchi, W. & Johnson, J. (1978). Types of organizational control and their relationship to emotional well-being, *Administrative Science Quarterly* **23**: 293–317.

Ouchi, W. & Maguire, M. (1975). Organizational control: Two functions, *Administrative Science Quarterly* **20**: 559–569.

Panagopoulos, A. (2009). Revisiting the link between knowledge spillovers and growth: An intellectual property perspective, *Economics of Innovation and New Technology* **18**(6): 533–546.

Parker, G. & Van Alstyne, M. (2005). Two-sided network effects: A theory of information product design, *Management Science* **51**(10): 1494–1504.

Parker, G. & Van Alstyne, M. (2008). Innovation, openness, and platform control, *Working Paper*. Available at SSRN: http://ssrn.com/abstract=1079712 .

Parnas, D. (1972). On the criteria to be used in decomposing systems into modules, *Communications of the ACM* **15**(12): 1053–1058.

Patrakosol, B. & Olson, D. (2007). How interfirm collaboration benefits IT innovation, *Information & Management* **44**: 53–62.

Penrose, E. (1959). *The Theory of the Growth of the Firm*, Oxford University Press, Oxford, U.K.

Perens, B. (2005). The emerging economic paradigm of open source, *First Monday* .

Perr, J., Appleyard, M. & Sullivan, P. (2010). Open for business: Emerging business models in open source software, *International Journal of Technology Management* **52**(3/4): 432–456.

Peteraf, M. & Barney, J. (2003). Unravelling the resource-based tangle, *Managerial and Decision Economics* **24**(4): 309–323.

Petter, S., Straub, D. & Rai, A. (2007). Specifying formative constructs in IS research, *MIS Quarterly* **31**(4): 623–656.

Piccoli, G. & Ives, B. (2003). Trust and unintended effects of behavior control in virtual teams, *MIS Quarterly* **27**(3): 365–395.

Picot, A., Reichwald, R. & Wigand, R. (1998). *Die grenzenlose Unternehmung: Information, Organisation und Management*, Gabler, Wiesbaden, Germany.

Piller, F. & Walcher, D. (2006). Toolkits for idea competitions: A novel method to integrate users in new product development, *R&D Management* **36**(3): 307–318.

Pisano, G. (1990). The R&D boundaries of the firm: An empirical analysis, *Administrative Science Quarterly* **35**: 153–176.

Pisano, G. & Teece, D. (2007). How to capture value from innovation: Shaping intellectual property and industry architecture, *California Management Review* **50**(1): 278–296.

Podsakoff, P., MacKenzie, S., Lee, J.-Y. & Podsakoff, N. (2003). Common method biases in behavioral research: A critical review of the literature and recommended remedies, *Journal of Applied Psychology* **88**(5): 879–903.

Poole, M., Van de Ven, A., Dooley, K. & Holmes, M. (2000). *Organizational Change and Innovation Processes - Theory and Methods for Research*, Oxford University Press, Oxford, U.K.

Popper, K. (1934). *The Logic of Scientific Discovery*, Hutchinson, London, U.K.

Porter, M. (1980). *Competitive Strategy*, Free Press, New York, NY.

Porter, M. (1996). What is strategy?, *Harvard Business Review* **72**(11): 61–78.

Powell, W. (1990). Neither market nor hierarchy: Network forms of organization, *Research in Organizational Behavior* **12**: 295–336.

Prahalad, C. & Hamel, G. (1990). The core competence of the corporation, *Harvard Business Review* **66**(3): 79–91.

Preece, J. (2000). *Online Communities: Designing Usability, Supporting Sociability*, John Wiley & Sons, Chichester, NY.

Raisch, S. & Birkinshaw, J. (2008). Organizational ambidexterity: Antecedents, outcomes, and moderators, *Journal of Management* **34**: 375–409.

Raisch, S., Birkinshaw, J., Probst, G. & Tushman, M. (2009). Organizational ambidexterity: Balancing exploitation and exploration for sustained performance, *Organization Science* **20**(4): 685–695.

Raymond, E. (1998). The Cathedral and the Bazaar, URL, www.catb.org/~esr/writings/cathedral-bazaar/, last access: 8/25/2011.

Reagans, R. & McEvily, B. (2003). Network structure and knowledge transfer: The effects of cohesion and range, *Administrative Science Quarterly* **48**: 240–267.

Reichwald, R. & Piller, F. (2006). *Interaktive Wertschöpfung - Open Innovation, Individualisierung und neue Formen der Arbeitsteilung*, Gabler, Wiesbaden, Germany.

Remedios, R. & Boreham, N. (2004). Organizational learning and employees' intrinsic motivation, *Journal of Education and Work* **17**(2): 219–235.

Riehle, D. (2009). The commercial open source business model, *Proceedings of the 15th Americas Conference on Information Systems (AMCIS)*, San Francisco, CA.

Riehle, D. (2011a). Controlling and steering open source projects, *IEEE Computer* **44**(7).

Riehle, D. (2011b). The single vendor commercial open source business model, *Information Systems and e-Business Management* **forthcoming**.

Rivkin, J. & Siggelkow, N. (2003). Balancing search and stability: Interdependencies among elements of organizational design, *Management Science* **49**(3): 290–311.

Roberts, J., Hann, I.-H. & Slaughter, S. (2006). Understanding the motives, participation, and performance of open source software developers: A longitudinal study of the apache projects, *Management Science* **52**(7): 984–999.

Rohrbeck, R., Hölzle, K. & Gemünden, H. (2009). Opening up for competitive advantage - How Deutsche Telekom creates a open innovation ecosystem, *R&D Management* **39**(4): 420–430.

Romanelli, E. (1991). The evolution of new organizational forms, *Annual Review of Sociology* **17**: 79–103.

Romme, A. (2003). Making a difference: Organization as design, *Organization Science* **14**(5): 558–573.

Rossi Lamastra, C. (2009). Software innovativeness: A comparison between proprietary and free/open source solutions offered by Italian SMEs, *R&D Management* **39**(2): 153–169.

Rost, K. (2011). The strength of strong ties in the creation of innovation, *Research Policy* **40**(4): 588–604.

Rothaermel, F. & Alexandre, M. (2009). Ambidexterity in technology sourcing: The moderating role of absorptive capacity, *Organization Science* **20**(4): 759–780.

Rothaermel, F. & Hill, C. (2005). Technological discontinuities and complementary assets: A longitudinal study of industry and firm performance, *Organization Science* **16**(1): 52–70.

Rothwell, R. (1986). Innovation and re-innovation: A role for the user, *Journal of Marketing Research* **2**(2): 109–123.

Rudner, R. (1966). *Philosophy of Social Science*, Prentice- Hall, Englewood Cliffs, NJ.

Ruhanen, L., Scott, N., Ritchie, B. & Tkaczynski, A. (2010). Governance: A review and synthesis of the literature, *Tourism Review* **65**(4): 4–16.

Rustagi, S., King, W. & Kirsch, L. (2008). Predictors of formal control usage in IT outsourcing partnerships, *Information Systems Research* **19**(2): 126–143.

Ryan, R. & Deci, E. (2000). Intrinsic and extrinsic motivations: Classic definitions and new directions, *Contemporary Educational Psychology* **25**(7): 54–67.

Sakakibara, M. (2002). Formation of R&D consortia: Industry and company effects, *Strategic Management Journal* **23**(11): 1033–1050.

Sampson, R. (2005). Experience effects and collaborative returns in R&D alliances, *Strategic Management Journal* **26**(11): 1009–1031.

Samuels, W. (1998). Introduction, *in* W. Samuels (ed.), *The Founding of Institutional Economics*, Routledge, London, U.K., pp. xi–xii.

Sanchez, R. & Mahoney, J. (1996). Modularity, flexibility, and knowledge management in product and organization design, *Strategic Management Journal* **17**: 63–76.

Saur-Amaral, I. & Amaral, P. (2010). Contract innovation organisations in action: Doing collaborative new product development outside the firm, *International Journal of Technology Intelligence and Planning* **6**(1): 42–62.

Sawhney, M., Verona, G. & Prandelli, E. (2005). Collaborating to create: The internet as a platform for customer engagement in product innovation, *Journal of Interactive Marketing* **19**(4): 4–17.

Schaarschmidt, M., Bertram, M. & Von Kortzfleisch, H. (2011). Exposing differences of governance approaches in single and multi vendor open source software development, *in* M. Nüttgens, A. Gadatsch, K. Kautz, I. Schirmer & N. Blinn (eds), *Governance and Sustainability in IS, IFIP AICT 366, LNCS*, Springer, Berlin, Germany, pp. 16–28.

Schaarschmidt, M. & Kilian, T. (2011). Open innovation and customer integration: A study of barriers in the Telecommunication industry, *Paper, presented at 11th European Academy of Management Conference (EURAM)*, Tallinn, Estonia.

Schaarschmidt, M. & Von Kortzfleisch, H. (2009). Divide et impera! The role of firms in large OSS consortia, *Proceedings of the 15th Americas Conference on Information Systems (AMCIS)*, San Francisco, CA.

Schaarschmidt, M. & Von Kortzfleisch, H. (2010). Examining investment strategies of venture capitalists in open source software, *Paper, presented at 10th European Academy of Management Conference (EURAM)*, Rome, Italy.

Schaarschmidt, M., Von Kortzfleisch, H., Valcárcel, S. & Lindermann, N. (2011). Web 2.0 enabled employee collaboration in diverse SME networks: A CEO's perspective, *Proceedings of the 19th European Conference on Information Systems (ECIS)*, Helsinki, Finland.

Schilling, M. (2008). *Strategic Management of Technological Innovation*, 2nd edn, McGraw-Hill, New York, NY.

Schilling, M. & Phelps, C. (2007). Interfirm collaboration networks: The impact of large-scale network structure on firm innovation, *Management Science* **53**(7): 1113–1126.

Schneider, C. (2009). *Governancestrukturen in großen firmengetriebenen Open Source Projekten*, Master thesis, University of Koblenz-Landau, Institute for Management, Koblenz, Germany.

Scholl, W. (1999). Restrictive control and information pathologies in organizations, *Journal of Social Issues* **55**(1).

Schubert, P. & Hampe, F. (2006). Mobile communities: How viable are their business models? An exemplary investigation of the leisure industry, *Electronic Commerce Research* **6**: 103–121.

Schumpeter, J. (1936). *The Theory of Economic Development: An Inquiry into Profits, Capital, Credit, Interest, and the Business Cycle*, Harvard University Press, Cambridge, MA.

Scott, W. (2008). *Institutions and Organization: Interests and Ideas*, 3rd edn, Sage Publications, Thousand Oaks, CA.

Scozzi, B., Crowston, K., Eseryel, Y. & Li, Q. (2008). Shared mental models among open source software developers, *Proceedings of the 41st Hawaii International Conference on System Science (HICSS-41)*, Big Island, Hawaii.

Seth, A. & Zinkhan, G. (1991). Strategy and the research process: A comment, *Strategic Management Journal* **12**(1): 75–82.

Setia, P., Rajagopal, B., Sambamurthy, V. & Calantone, R. (2011). How peripheral developers contribute to open-source software development, *Information Systems Research* **in press**.

Settle, T. (1979). Popper on 'when is a science not a science'?, *Systematic Zoology* **28**(4): 521–529.

Shah, S. (2000). Sources and patterns of innovation in a consumer products field: Innovations in sporting equipment, *MIT Sloan School of Management, Working Paper No. 4105* .

Shah, S. (2006). Motivation, governance, and the viability of hybrid forms in open source software development, *Management Science* **52**(7): 1000–1014.

Shah, S. & Tripsas, M. (2007). The accidental entrepreneur: The emergent and collective process of user entrepreneurship, *Strategic Entrepreneurship Journal* **1**: 123–140.

Shaikh, M. & Cornford, T. (2003). Version management tools: CVS to BK in the Linux kernel, *Proceedings of the 25th International Conference on Software Engineering*, Portland, Oregon, pp. 127–132.

Shane, S. & Ulrich, K. (2004). Technological innovation, product development, and entrepreneurship in management science, *Management Science* **5**(2): 133–144.

Shavell, S. & Van Ypersele, T. (2001). Rewards versus intellectual property rights, *Journal of Law and Economics* **44**(2): 525–547.

Shaw, M. (1964). Communication networks, *in* L. Berkowitz (ed.), *Advances in Experimental Social Psychology, Vol. 1*, Academic, New York, NY, pp. 111–147.

Sidhu, J., Commandeur, H. & Volberda, H. (2007). The multifaceted nature of exploration and exploitation: Value of supply, demand, and spatial search for innovation, *Organization Science* **18**(1): 20–38.

Siggelkow, N. & Rivkin, J. (2006). When exploration backfires: Unintended consequences of multilevel organizational search, *Academy of Management Journal* **49**(4): 779–795.

Siggelkow, N. & Rivkin, J. (2009). Hiding the evidence of valid theories: How coupled search processes obscure performance differences among organizations, *Administrative Science Quarterly* **54**(4): 602–634.

Silverman, B., Nickerson, J. & Freeman, J. (1997). Profitability, transactional alignment, and organizational mortality in the us trucking industry, *Strategic Management Journal* **18**: 31–52.

Simonin, B. (2004). An empirical investigation of the process of knowledge transfer in international strategic alliances, *Journal of International Business Studies* **35**(5): 407–427.

Singh, A. & Soltani, E. (2010). Knowledge management practices in Indian information technology companies, *Total Quality Management* **21**(2): 145–157.

Singh, J. (2008). Distributed R&D, cross-regional knowledge integration and quality of innovative output, *Research Policy* **37**: 77–96.

Singh, P. & Tan, Y. (2010). Developer heterogeneity and formation of communication networks in open source software projects, *Journal of Management Information Systems* **27**(3): 179–210.

Sleeswijk-Visser, F., Van der Lugt, R. & Stappers, P. (2007). Sharing user experience in the product innovation process: Participatory design need participatory communication, *Creativity and Innovation Management* **16**(1): 35–45.

Snell, S. (1992). Control theory in strategic human resource management: The mediating effect of administrative information, *Academy of Management Journal* **35**(2): 292–327.

Snow, D., Soule, S. & Kriesi, H. (2004). Mapping the terrain, *in* D. Snow, S. Soule & H. Kriesi (eds), *The Blackwell Companion to Social Movements*, Wiley-Blackwell, Malden, MA, pp. 3–16.

Sojer, M. (2011). *Reusing Open Source Code: Value Creation and Value Appropriation Perspectives on Knowledge Reuse*, Gabler, Wiesbaden, Germany.

Sojer, M. & Henkel, J. (2010). Code reuse in open source software development: Quantitative evidence, drivers, and impediments, *Journal of the Association for Information Systems* **11**(12): 868–901.

Somaya, D., Teece, D. & Wakeman, S. (2011). Innovation in multi-invention contexts: Mapping solutions to technological and intellectual property complexity, *California Management Review* **53**(4): 47–79.

Sosa, M., Eppinger, S. & Rowles, C. (2004). The misalignment of product architecture and organizational structure in complex product development, *Management Science* **50**(12): 1674–1689.

Sparrowe, R., Liden, R., Wayne, S. & Kraimer, M. (2001). Social networks and the performance of individuals and groups, *Academy of Management Journal* **44**(2): 316–325.

Spencer, M. (1987). The imperfect empiricism of the social sciences, *Sociological Forum* **2**(2): 331–372.

St. Laurent, A. (2004). *Understanding Open Source and Free Software Licensing*, O'Reilly Media, Cambridge, MA.

Stevens, J. (2002). *Applied Multivariate Statistics for the Social Sciences*, Lawrence Erblaum, Mahwah, NJ.

Stevenson, H. & Jarillo, J. (1990). A paradigm of entrepreneurship: Entrepreneurial management, *Strategic Management Journal* **11**: 17–27.

Stewart, K., Ammeter, A. & Maruping, L. (2006). Impacts of license choice and organizational sponsorship on user interest and development activity in open source software projects, *Information Systems Research* **17**(2): 126–144.

Stewart, K. & Gosain, S. (2006). The impact of ideology on effectiveness in open source software development teams, *MIS Quarterly* **30**(2): 291–314.

Straub, D., Boudreau, M.-C. & Gefen, D. (2004). Validation guidelines for IS positivist research, *Communications of the Association for Information Systems* **13**(1): 380–427.

Störig, H.-J. (1968). *Kleine Weltgeschichte der Philosophie*, Kohlhammer, Stuttgart.

Stürmer, M., Spaeth, S. & Von Krogh, G. (2009). Extending private-collective innovation: A case study, *R&D Management* **39**(2): 170–191.

Stuckenschmidt, H. & Klein, M. (2004). Structure-based partitioning of large concept hierarchies, *in* S. McIlraith, D. Plexousakis & F. Van Harmelen (eds), *Proceedings of the Third International Semantic Web Conference (ISWC 2004)*, Hiroshima, Japan, p. 289–303.

Suddaby, R. (2010). Challenges for institutional theory, *Journal of Management Inquiry* **19**(1): 14–20.

Sullivan, A. & Sheffrin, S. (2003). *Economics: Principles in Action*, Pearson Prentice Hall, Upper Saddle River, NJ.

Sundaramurthy, C. & Lewis, M. (2003). Control and collaboration: Paradoxes of governance, *Academy of Management Review* **28**(3): 447–465.

Tabachnik, B. & Fidell, L. (1989). *Using Multivariate Statistics*, 2nd edn, Harper Collins, New York, NY.

Tassoul, M. (2006). *Creativity Facilitation*, VSSD, Delft, NL.

Teece, D. (1986). Profiting from technological innovation: Implications for integration, collaboration, licensing and public policy, *Research Policy* **15**: 285–305.

Teece, D. (2000). *Managing Intellectual Capital*, Oxford University Press, Oxford, U.K.

Teece, D. (2003). *Essays in Technology Management and Policy*, World Scientific Publishing, London, U.K.

Teece, D. (2009). *Dynamic Capabilities and Strategic Management: Organizing for Innovation and Growth*, Oxford University Press, Oxford, U.K.

Teece, D. (2010a). Alfred Chandler and "capabilities" theories of strategy and management, *Industrial and Corporate Change* **19**(2): 297–316.

Teece, D. (2010b). Business models, business strategy, and innovation, *Long Range Planning* **43**(2/3): 172–194.

Thiel, C. (1984). Konstruktivismus, *in* J. Mittelstrass (ed.), *Enzyklopädie Philosophie und Wissenschaftstheorie*, Metzler, Mannheim, pp. 449–453.

Thomke, S. & Von Hippel, E. (2002). Customers as innovators: A new way to create value, *Harvard Business Review* **80**(4): 74–81.

Timmers, P. (1998). Business models for electronic markets, *Electronic Markets* **8**(2): 3–8.

Torkar, R., Minoves, P. & Garrigós (2011). Adopting free/libre/open source software practices, techniques, and methods for industrial use, *Journal of the Association for Information Systems* **12**(1): 88–122.

Toutenburg, H. & Heumann, C. (2008). *Induktive Statistik - Eine Einführung mit R und SPSS*, 4th edn, Springer, Berlin, Germany.

Tripsas, M. (1997). Unraveling the process of creative destruction: Complementary assets and incumbent survival in the typesetter industry, *Strategic Management Journal* **18**: 119–142.

Trott, P. & Hartmann, D. (2009). Why 'open innovation' is old wine in new bottles, *International Journal of Innovation Management* **13**(4): 715–736.

Tsai, W. (2001). Knowledge transfer in intraorganizational networks – Effects of network position and absorptive capacity on business unit innovation and performance, *Academy of Management Journal* **44**(5): 996–1004.

Tuomi, I. (2001). Internet, innovation, and open source: Actors in the network, *First Monday* **6**(1).

Tuomi, I. (2004). Evolution of the Linux credits file: Methodological challenges and reference data for open source research, *First Monday* **9**(6).

Tuppuraa, A., Hurmelinna-Laukkanena, P., Puumalainena, K. & Jantunena, A. (2010). The influence of appropriability conditions on the firm's entry timing orientation, *Journal of High Technology Management Research* **21**(2): 97–107.

Tushman, M. & Scanlan, T. (1981a). Boundary spanning individuals: Their role in information transfer and their antecedents, *Academy of Management Journal* **24**(2): 289–305.

Tushman, M. & Scanlan, T. (1981b). Characteristics and external orientations of boundary spanning individuals, *Academy of Management Journal* **24**(1): 83–98.

Ulrich, K. & Eppinger, S. (2004). *Product Design and Development*, 3rd edn, MacGraw-Hill, New York, NY.

Uzzi, B. (1997). Social structure and competition in interfirm networks: The paradox of embeddedness, *Administrative Science Quarterly* **42**: 35–67.

Uzzi, B. & Lancaster, R. (2003). The role of relationships in interfirm knowledge transfer and learning: The case of corporate debt markets, *Management Science* **49**: 383–399.

Valcárcel, S. (2002). *Theorie der Unternehmung und Corporate Governance: Eine vertrags- und ressourcenbezogene Betrachtung*, Deutscher Universitätsverlag, Wiesbaden, Germany.

Valverde, S. & Solé, R. (2006). Self-organization and hieracrchy in open source social networks, *Santa Fe Institute Research Paper APS/123-QED* .

Van Aken, J. (2004). Management research based on the paradigm of the design sciences: The quest for field-tested and grounded technological rules, *Journal of Management Studies* **41**(2): 219–246.

Van de Ven, A. & Garud, R. (1993). A community perspective on the emergence of innovations, *Journal of Engineering and Technology Management* **10**(1-2): 23–51.

Van de Vrande, V., De Jong, J., Vanhaverbeke, W. & De Rochemont, M. (2009). Open innovation in SMEs: Trends, motives, and management challenges, *Technovation* **29**(6/7): 423–437.

Van de Vrande, V., Lemmens, C. & Vanhaverbeke, W. (2006). Choosing governance modes for external technology sourcing, *R&D Management* **36**(3): 347–363.

Van de Vrande, V., Vanhaverbeke, W. & Duysters, G. (2009). External technology sourcing: The effect of uncertainty on governance mode choice, *Journal of Business Venturing* **24**: 62–80.

Van de Vrande, V., Vanhaverbeke, W. & Gassmann, O. (2010). Broadening the scope of open innovation: Past research, current state and future directions, *International Journal of Technology Management* **52**(3/4): 221–235.

Van Dijk, C. & Van den Ende, J. (2002). Suggestion systems: Transferring employee creativity into practicable ideas, *R&D Management* **32**(5): 387–395.

Van Fraassen, B. (1980). *The Scientific Image*, Claredon Press, Oxford, U.K.

Van Fraassen, B. (2008). *Scientific Representation: Paradoxes of Perspective*, Oxford University Press, Oxford, U.K.

Van Schewick, B. (2005). *Architecture and Innovation: The Role of the End-to-End Arguments in the Original Internet*, PhD thesis, TU Berlin, Berlin, Germany.

Van Sell, M., Brief, A. & Schuler, R. (1981). Role conflict and role ambiguity: Integration of the literature and directions for future research, *Human Relations* **34**: 43–71.

Vanhaverbeke, W., Cloodt, M. & van de Vrande, V. (2008). Connecting absorptive capacity and open innovation, URL, http://www.cas.uio.no/research/0708innovation/CASworkshop_VanhaverbekeEtAl.pdf, last access: 9/14/2010.

Vanhaverbeke, W., Duysters, G. & Noorderhaven, N. (2002). External technology sourcing through alliances or acquisitions: An analysis of the application-specific integrated circuits industry, *Organization Science* **13**(6): 714–733.

Vargo, L., Maglio, P. & Akaka, M. (2008). On value and value co-creation: A service systems and service logic perspective, *European Management Journal* **26**: 145–152.

Vargo, S. & Akaka, M. (2009). Service-dominant logic as a foundation for service science: Clarifications, *Service Science* **1**(1): 32–41.

Vargo, S. & Lusch, R. (2004). Evolving to a new dominant logic for marketing, *Journal of Marketing* **68**: 1–17.

Vargo, S. & Lusch, R. (2008). Service-dominant logic: Continuing the evolution, *Journal of the Academy of Marketing Science* **36**(1): 1–10.

Ven, K. & De Bruyn, P. (2011). Factors affecting the development of absorptive capacity in the adoption of open source software, *International Journal of Open Source Software & Processes* **3**(1): 17–38.

Ven, K. & Verelst, J. (2008). The impact of ideology on the organizational adoption of open source software, *Journal of Database Management* **19**(2): 58–72.

Verona, G. (1999). A resource-based view of product development, *Academy of Management Review* **24**(1): 132–142.

Villalonga, B. & McGahan, A. (2005). The choice among acquisitions, alliances and divestures, *Strategic Management Journal* **26**: 1183–1208.

Von Glasersfeld, E. (1995). *Radical Constructivism. A Way of Knowing and Learning*, Falmer Press, London, U.K.

Von Hippel, E. (1978a). A customer active paradigm for industrial product idea generation, *Research Policy* **7**(3): 240–266.

Von Hippel, E. (1978b). Successful industrial products from customer ideas, *Journal of Marketing* **42**(1): 39–49.

Von Hippel, E. (1986). Lead users: A source of novel product concepts, *Management Science* **32**(7): 791–805.

Von Hippel, E. (1988). *Sources of Innovation*, MIT Press, Cambridge, MA.

Von Hippel, E. (1990). Task partitioning: An innovative process variable, *Research Policy* **19**: 407–418.

Von Hippel, E. (1994). Sticky information and the locus of problem solving, *Management Science* **40**: 429–439.

Von Hippel, E. (2005). *Democratizing Innovation*, MIT Press, Cambridge, MA.

Von Hippel, E. (2007). Horizontal innovation networks – by and for users, *Industrial and Corporate Change* **16**: 293–315.

Von Hippel, E. & Katz, R. (2002). Shifting innovation to users via toolkits, *Management Science* **48**(7): 821–833.

Von Hippel, E. & Von Krogh, G. (2003). Open source software and the private-collective innovation model: Issues for organization science, *Organization Science* **14**: 209–225.

Von Kortzfleisch, H. (2004). *Organisatorische Balancierung von Informations- und Kommunikationstechnologien*, Eul Verlag, Lohmar, Germany.

Von Kortzfleisch, H. & Mergel, I. (2002). Getting over 'knowledge is power': Incentive systems for knowledge management in business consulting companies, *in* D. White (ed.), *Knowledge Mapping and Management*, IRM Press, Hershey, pp. 244–253.

Von Kortzfleisch, H., Schaarschmidt, M. & Magin, P. (2010). Open scientific entrepreneurship: How the open source paradigm can foster entrepreneurial activities in scientific institutions, *International Journal of Open Source Software & Processes* **2**(4): 48–66.

Von Krogh, G. & Spaeth, S. (2007). The open source software phenomenon: Characteristics that promote research, *Journal of Strategic Information Systems* **16**: 236–253.

Von Krogh, G., Spaeth, S. & Lakhani, K. (2003). Community, joining, and specialization in open source software innovtion: A case study, *Research Policy* **32**: 1217–1241.

Von Krogh, G. und Haeflinger, S. (2010). Opening up design science: The challange of designing for reuse and joint development, *Journal of Strategic Information Systems* **19**(4): 232–241.

Wade, M. & Hulland, J. (2004). Review: The resource-based view and information system research: Review, extension, and suggestions for future research, *MIS Quarterly* **28**(1): 107–142.

Wagstrom, P. (2009). *Vertical Interaction in Open Source Engineering Communities*, PhD thesis, Carnegie Mellon University, Carnegie Institute of Technology and School of Computer Science, Pittsburgh, PA.

Wallace, J. (1995). Organizational and professional commitment in professional and nonprofessional organizations, *Administrative Science Quarterly* **40**(2): 228–255.

Walsh, G. & Beatty, S. (2007). Customer-based corporate reputation of a service firm: Scale development and validation, *Journal of the Academy of Marketing Science* **35**(1): 127–143.

Walsh, G., Klee, A. & Kilian, T. (2009). *Marketing – Eine Einführung auf der Grundlage von Case Studies*, Springer, Berlin, Germany.

Walsh, G., Schaarschmidt, M. & Von Kortzfleisch, H. (2011). Harnessing free external resources: Evidence from the open source field, *unpublished working paper* .

Wasko, M. & Faraj, S. (2000). It's what one does: Why people participate and help others in electronic communities of practice, *Journal of Strategic Information Systems* **9**(2-3): 155–173.

Wasko, M. & Faraj, S. (2005). Why should I share? Examining social software and knowledge contribution in electronic networks of practice, *MIS Quarterly* **29**: 35–57.

Wasserman, S. & Faust, K. (1994). *Social Network Analysis – Methods and Applications*, Cambridge University Press, New York, NY.

Wassmer, U. & Dussauge, P. (2011). Value creation in alliance portfolios: The benefits and costs of network resource interdependencies, *European Management Review* **8**: 47–64.

Watson, R., Boudreau, M.-C., York, P., Greiner, M. & Wynn, D. (2008). The business of open source, *Communications of the ACM* **51**(4): 41–46.

Weber, S. (2005). Patterns of governance in open source, *Open Sources 2.0: The Continuing Evolution*, O'Reilly, Sebastopol, CA, pp. 361–372.

Weigelt, K. & Camerer, C. (1988). Reputation and corporate strategy: A review of recent theory and applications, *Strategic Management Journal* **9**: 443–454.

Weinmann, M. (2009). *Diffusion von Meinungen in sozialen Netzwerken des Web 2.0*, GRIN Verlag, München, Germany.

Wernerfelt, B. (1984). A resource-based view of the firm, *Strategic Management Journal* **5**(2): 171–180.

Wesselius, J. (2008). The bazaar inside the cathedral: Business models for internal markets, *IEEE Software* **25**(3): 60–66.

West, J. (2003). How open is open enough? Melding proprietary and open source platform strategies, *Research Policy* **32**: 1259–1285.

West, J. (2007). Value capture and value networks in open source vendor strategies, *Proceedings of the 40th Hawaii International Conference on System Sciences (HICSS-40)*, IEEE Computer Society, Waikoloa, Hawaii.

West, J. & Bogers, M. (2010). Contrasting innovation creation and commercialization within open, user and cumulative innovation, *Proceedings of the Sixty-Ninth Annual Meeting of the Academy of Management*, Montreal, Canada.

West, J. & Dedrick, J. (2001). Open source standardization: The rise of Linux in the network era, *Knowledge, Technology, & Policy* **14**(2): 88–112.

West, J. & Gallagher, S. (2006). Challenges of open innovation: The paradox of firm investment in open source software, *R&D Management* **36**(3): 319–331.

West, J. & Lakhani, K. (2008). Getting clear about communities in open innovation, *Industry and Innovation* **15**(2): 223–231.

West, J. & O'Mahony, S. (2008). The role of participation architecture in growing sponsored open source communities, *Industry and Innovation* **15**(2): 145–168.

White, H. (1981). Where do markets come from?, *American Journal of Sociology* **87**(3): 517–547.

Williamson, O. (1975). *Markets and Hierarchies - Analysis and Antitrust Implications*, The Free Press, New York, NY.

Williamson, O. (1991). Comparative economic organization: The analysis of discrete structural alternatives, *Administrative Science Quarterly* **36**(2): 269–296.

Williamson, O. (1996). Transaction cost economics and the carnegie connection, *in* O. Williamson (ed.), *The Mechanisms of Governance*, Oxford University Press, Oxford, U.K., pp. 23–29.

Winter, R. (2008). Design science research in Europe, *European Journal of Information Systems* **17**: 470–475.

Wirtz, B. (2008). *Electronic Business*, 3rd edn, Gabler, Wiesbaden, Germany.

Wu, C.-G., Gerlach, J. & Young, C. (2007). An empirical analysis of open source software developers? Motivations and continuance intentions, *Information & Management* **44**: 253–262.

Xu, B., Jones, D. & Shao, B. (2009). Volunteers' involvement in online community based software development, *Information & Management* **46**: 151–158.

Xu, J., Christley, S. & Madey, G. (2006). Application of social network analysis to the study of open source software, *in* J. Bitzer & J. Schröder (eds), *The Economics of Open Source Software*, Elsevier, Amsterdam, pp. 247–270.

Yehezkel, O. & Lerner, M. (2009). Born to be wild? On managerial capabilities and business performance in technology start-ups, *International Studies of Management and Organization* **39**(3): 6–31.

Yli-Renko, H., Autio, E. & Sapienza, H. (2001). Social capital, knowledge acquisitions, and knowledge exploitation in young technology-based firms, *Strategic Management Journal* **22**(6/7): 587–613.

Young, M. (1994). *The Rise of Meritocracy*, Transaction Publishers, Piscataway, NJ.

Zahra, S. & George, G. (2002). Absorptive capacity: A review, reconceptualization, and extension, *Academy of Management Review* **27**(2): 185–203.

Zhang, H., Shu, C., Jiang, X. & Malter, A. (2010). Managing knowledge for innovation: The role of cooperation, competition, and alliance nationality, *Journal of International Marketing* **18**(4): 74–94.

Zirn, F. (2004). *Softwareschutz zwischen Urheberrecht und Patentrecht. Aktuelle Entwicklungen vor Historischem Hintergrund und Internationaler Zusammenhang*, Ibidem-Verlag, Stuttgart, Germany.